Enzyme Reaction Engineering

Enzyme Reaction Engineering

Editor

Dr. T.P. Jayadev Reddy

BIOTECH

2007

Biotech Books

Delhi - 110 035

ISBN 81-7622-168-6

Published by : **BIOTECH BOOKS**
1123/74, Tri Nagar,
DELHI – 110 035
Phone: 27382765
e-mail: biotechbooks@yahoo.co.in

Showroom : 4762-63/23, Ansari Road, Darya Ganj,
NEW DELHI - 110 002
Phone: 23245578, 23244987

Printed at : **Chawla Offset Printers**
New Delhi – 110 052

PRINTED IN INDIA

Preface

Enzyme reaction engineering has always been an interdisciplinary subject with applications ranging from biocatalysts preparation and characterisation, to the use of biocatalysts in various biotransformations at different process conditions such as reaction medium reactant and product solubility, reaction kinetics etc.

A most exciting development over the last few years is the application genetic engineering techniques to enzyme technology. There are a number of properties which may be improved or altered by genetic engineering including the yield and kinetics of the enzyme, the ease of downstream processing and various safety aspects. Enzymes from dangerous or unapproved microorganisms and from slow growing or limited plant or animal tissue may be cloned into safe high-production microorganisms. In the future, enzymes may be redesigned to fit more appropriately into industrial processes.

This book describes the recent developments in enzyme reaction engineering based on modern sciences and technology. It builds a bridge between basic knowledge and biotechnology in the field of biochemistry and enzymology. The book will provide the reader with the most up-to-date information on development of enzyme reaction engineering.

Editor

Contents

Chapter 1

Enzymes: Nature and Functions

Enzymes are biological catalysts responsible for supporting almost all of the chemical reactions that maintain animal homeostasis. Because of their role in maintaining life processes, the assay and pharmacological regulation of enzymes have become key elements in clinical diagnosis and therapeutics. The macromolecular components of almost all enzymes are composed of protein, except for a class of RNA modifying catalysts known as ribozymes. Ribozymes are molecules of ribonucleic acid that catalyze reactions on the phosphodiester bond of other RNAs.

Enzymes are found in all tissues and fluids of the body. Intracellular enzymes catalyze the reactions of metabolic pathways. Plasma membrane enzymes regulate catalysis within cells in response to extracellular signals, and enzymes of the circulatory system are responsible for regulating the clotting of blood. Almost every significant life process is dependent on enzyme activity.

System of Enzyme Classifications

Traditionally, enzymes were simply assigned names by the investigator who discovered the enzyme. As knowledge expanded, systems of enzyme classification became more comprehensive and complex. Currently enzymes are grouped into six functional classes

1. Oxidoreductases Act on many chemical groupings to add or remove hydrogen atoms.
2. Transferases Transfer functional groups between donor and acceptor molecules. Kinases are specialized transferases that regulate metabolism by transferring phosphate from ATP to other molecules.
3. Hydrolases Add water across a bond, hydrolyzing it.
4. Lyases Add water, ammonia or carbon dioxide across double bonds, or remove these elements to produce double bonds.
5. Isomerases Carry out many kinds of isomerization: L to D isomerizations, mutase reactions (shifts of chemical groups) and others.
6. Ligases Catalyze reactions in which two chemical groups are joined (or ligated) with the use of energy from ATP.

Enzymes are also classified on the basis of their composition. Enzymes composed wholly of protein are known as simple enzymes in contrast to complex enzymes, which are composed of protein plus a relatively small organic molecule. Complex enzymes are also known as holoenzymes. In this terminology the protein component is known as the apoenzyme, while the non-protein component is known as the coenzyme or prosthetic group where prosthetic group describes a complex in which the small organic molecule is bound to the apoenzyme by covalent bonds; when the binding between the apoenzyme and non-protein components is non-covalent, the small organic molecule is called a coenzyme.

Many prosthetic groups and coenzymes are water-soluble derivatives of vitamins. It should be noted that the main clinical symptoms of dietary vitamin insufficiency generally arise from the malfunction of enzymes, which lack sufficient cofactors derived from vitamins to maintain homeostasis.

The non-protein component of an enzyme may be as simple as a metal ion or as complex as a small non-protein organic molecule. Enzymes that require a metal in their composition are known as metalloenzymes if they bind and retain their metal atom(s) under all conditions, that is with very high affinity. Those which have a lower affinity for metal ion, but still require the metal ion for activity, are known as metal-activated enzymes.

Coenzymes

The functional role of coenzymes is to act as transporters of chemical groups from one reactant to another. The chemical groups carried can be as simple as the hydride ion ($H^+ + 2e^-$) carried by NAD or the mole of hydrogen carried by FAD; or they can be even more complex than the amine ($-NH_2$) carried by pyridoxal phosphate. Since coenzymes are chemically changed as a consequence of enzyme action, it is often useful to consider coenzymes to be a special class of substrates, or second substrates, which are common to many different holoenzymes. In all cases, the coenzymes donate the carried chemical grouping to an acceptor molecule and are thus regenerated to their original form. This regeneration of coenzyme and holoenzyme fulfills the definition of an enzyme as a chemical catalyst, since (unlike the usual substrates, which are used up during the course of a reaction) coenzymes are generally regenerated.

Substrate Type

Although enzymes are highly specific for the kind of reaction they catalyze, the same is not always true of substrates they attack. For example, while succinic dehydrogenase (SDH) always catalyzes an oxidation-reduction reaction and its substrate is invariably succinic acid, alcohol dehydrogenase (ADH) always catalyzes oxidation-reduction reactions but attacks a number of different alcohols, ranging from methanol to butanol. Generally, enzymes having broad substrate specificity are most active against one particular substrate. In the case of ADH, ethanol is the preferred substrate.

Enzymes also are generally specific for a particular steric configuration (optical isomer) of a substrate. Enzymes that attack D sugars will not attack the corresponding L isomer. Enzymes that act on L amino acids will not employ the corresponding D optical isomer as a substrate. The enzymes known as racemases provide a striking exception to these generalities; in fact, the role of racemases is to convert D isomers to L isomers and vice versa. Thus racemases attack both D and L forms of their substrate.

As enzymes have a more or less broad range of substrate specificity, it follows that a given substrate may be acted on by a number of different enzymes, each of which uses the same substrate(s) and produces the same product(s). The individual members of a set of enzymes sharing such characteristics are known as isozymes. These are the products of genes that vary only slightly; often, various isozymes of a group are expressed in different tissues of the body. The best studied set of isozymes is the lactate dehydrogenase (LDH) system. LDH is a tetrameric enzyme composed of all possible arrangements of two different protein subunits; the subunits are known as H (for heart) and M (for skeletal muscle). These subunits combine in various combinations leading to 5 distinct isozymes. The all H isozyme is characteristic of that from heart tissue, and the all M isozyme is typically found in skeletal muscle and liver. These isozymes all catalyze the same chemical reaction, but they exhibit differing degrees of efficiency. The detection of specific LDH isozymes in the blood is highly diagnostic of tissue damage such as occurs during cardiac infarct.

Substrate Interactions

The favored model of enzyme substrate interaction is known as the induced fit model. This model proposes that the initial interaction between enzyme and substrate is relatively weak, but that these weak interactions rapidly induce conformational changes in the enzyme that strengthen binding and bring catalytic sites close to substrate bonds to be altered. After binding takes place, one or more mechanisms of catalysis generates transition- state complexes and reaction products. The possible mechanisms of catalysis are four in number:

1. *Catalysis by Bond Strain*: In this form of catalysis, the induced structural rearrangements that take place with the binding of substrate and enzyme ultimately produce strained substrate bonds, which more easily attain the transition state. The new conformation often forces substrate atoms and bulky catalytic groups, such as aspartate and glutamate, into conformations that strain existing substrate bonds.

2. *Catalysis by Proximity and Orientation*: Enzyme-substrate interactions orient reactive groups and bring them into proximity with one another. In addition to inducing strain, groups such as aspartate are frequently chemically reactive as well, and their proximity and orientation toward the substrate thus favors their participation in catalysis.

3. *Catalysis Involving Proton Donors (Acids) and Acceptors (Bases)*: Other mechanisms also contribute significantly to the completion of catalytic events initiated by a strain mechanism, for example, the use of glutamate as a general acid catalyst (proton donor).

4. *Covalent Catalysis*: In catalysis that takes place by covalent mechanisms, the substrate is oriented to active sites on the enzymes in such a way that a covalent intermediate forms between the enzyme or coenzyme and the substrate. One of the best-known examples of this mechanism is that involving proteolysis by serine proteases, which include both digestive enzymes (trypsin, chymotrypsin, and elastase) and several enzymes of the blood clotting cascade. These proteases contain an active site serine whose R group hydroxyl forms a covalent bond with a carbonyl carbon of a peptide bond, thereby causing hydrolysis of the peptide bond.

Enzymes as Biological Catalysts

In cells and organisms most reactions are catalyzed by enzymes, which are regenerated during the course of a reaction. These biological catalysts are physiologically important because they speed up the rates of reactions that would otherwise be too slow to support life. Enzymes increase reaction rates—sometimes by as much as one millionfold, but more typically by about one thousand fold. Catalysts speed up the forward and reverse reactions proportionately so that, although the magnitude of the rate constants of the forward and reverse reactions is are increased, the ratio of the rate constants remains the same in the presence or absence of enzyme.

Since the equilibrium constant is equal to a ratio of rate constants, it is apparent that enzymes and other catalysts have no effect on the equilibrium constant of the reactions they catalyze. Enzymes increase reaction rates by decreasing the amount of energy required to form a complex of reactants that is competent to produce reaction products. This complex is known as the activated state or transition state complex for the reaction. Enzymes and other catalysts accelerate reactions by lowering the energy of the transition state. The free energy required to form an activated complex is much lower in the catalyzed reaction. The amount of energy required to achieve the transition state is lowered; consequently, at any instant a greater proportion of the molecules in the population can achieve the transition state. The result is that the reaction rate is increased.

Michaelis-Menton Kinetics

In typical enzyme-catalyzed reactions, reactant and product concentrations are usually hundreds or thousands of times greater than the enzyme concentration. Consequently, each enzyme molecule catalyzes the conversion to product of many reactant molecules. In biochemical reactions, reactants are commonly known as substrates. The catalytic event that converts substrate to product involves the formation of a transition state, and it occurs most easily at a specific binding site on the enzyme. This site, called the catalytic site of the enzyme, has been evolutionarily structured to provide specific, high-affinity binding of substrate(s) and to provide an environment that favors the catalytic events. The complex that forms when substrate(s) and enzyme combine is called the enzyme substrate (ES) complex. Reaction products arise when the ES complex breaks down releasing free enzyme.

Between the binding of substrate to enzyme, and the reappearance of free enzyme and product, a series of complex events must take place. At a minimum an ES complex must be formed; this complex must pass to the transition state (ES*); and the transition state complex must advance to an enzyme product complex (EP). The latter is finally competent to dissociate to product and free enzyme. The series of events can be shown thus:

E + S <---> ES <---> ES* <---> EP <---> E + P

The kinetics of simple reactions like that above were first characterized by biochemists Michaelis and Menten. The concepts underlying their analysis of enzyme kinetics continue to provide the cornerstone for understanding metabolism today, and for the development and clinical use of drugs aimed at selectively altering rate constants and interfering with the progress of disease states. The Michaelis-Menten equation:

$$v_1 = \frac{V_{max}[S]}{\{K_m + [S]\}}$$

is a quantitative description of the relationship among the rate of an enzyme- catalyzed reaction [v_1], the concentration of substrate [S] and two constants, V_{max} and K_m (which are set by the particular equation). The symbols used in the Michaelis-Menton equation refer to the reaction rate [v_1], maximum reaction rate (V_{max}), substrate concentration [S] and the Michaelis-Menton constant (K_m).

The Michaelis-Menten equation can be used to demonstrate that at the substrate concentration that produces exactly half of the maximum reaction rate, i.e.,1/2 V_{max}, the substrate concentration is numerically equal to K_m. This fact provides a simple yet powerful bioanalytical tool that has been used to characterize both normal and altered enzymes, such as those that produce the symptoms of genetic diseases. Rearranging the Michaelis-Menton equation leads to:

$$K_m = [S]\left\{\left[\frac{V_{max}}{V_1}\right]-1\right\}$$

From this equation it should be apparent that when the substrate concentration is half that required to support the maximum rate of reaction, the observed rate, v1, will, be equal to V_{max} divided by 2; in other words, v_1 = [V_{max}/2]. At this substrate concentration V_{max}/v_1 will be exactly equal to 2, with the result that

$$[S](1) = K_m$$

The latter is an algebraic statement of the fact that, for enzymes of the Michaelis-Menten type, when the observed reaction rate is half of the maximum possible reaction rate, the substrate concentration is numerically equal to the Michaelis-Menten constant. In this derivation, the units of K_m are those used to specify the concentration of S, usually Molarity.

The Michaelis-Menten equation has the same form as the equation for a rectangular hyperbola; graphical analysis of reaction rate (v) versus substrate concentration [S] produces a hyperbolic rate plot.

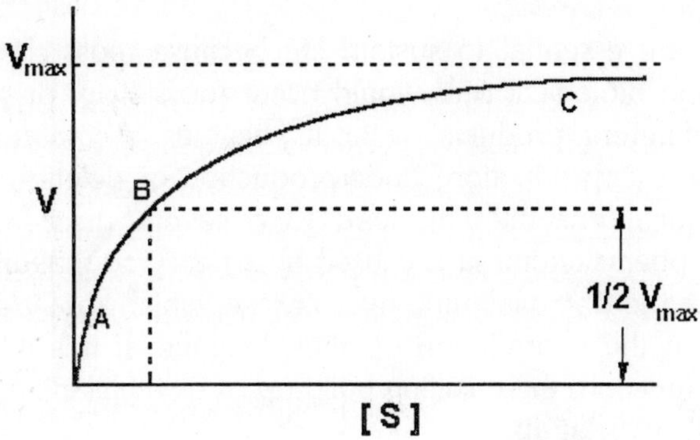

Figure 1: Graphic analysis of reaction rate vs. susbstrate concentration

The key features of the plot are marked by points A, B and C. At high substrate concentrations the rate represented by point C the rate of the reaction is almost equal to V_{max}, and the difference in rate at nearby concentrations of substrate is almost negligible. If the Michaelis-Menten plot is extrapolated to infinitely high substrate concentrations, the extrapolated rate is equal to V_{max}. When the reaction rate becomes independent of substrate concentration, or nearly so, the rate is said to be zero order.

The very small differences in reaction velocity at substrate concentrations around point C (near V_{max}) reflect the fact that at these concentrations almost all of the enzyme molecules are bound to substrate and the rate is virtually independent of substrate, hence zero order. At lower substrate concentrations, such as at points A and B, the lower reaction velocities indicate that at any moment only a portion of the enzyme molecules are bound to the substrate. In fact, at the substrate concentration denoted by point B, exactly half the enzyme molecules are in an ES complex at any instant and the rate is exactly one half of V_{max}. At substrate concentrations near point A the rate appears to be directly proportional to substrate concentration, and the reaction rate is said to be first order.

Enzymes and Reactions

Enzymes are essential to sustain life because most chemical reactions in biological cells would occur too slowly, or would lead to different products, without enzymes. A malfunction (mutation, overproduction, underproduction or deletion) of a single critical enzyme can lead to a severe disease. For example, phenylketonuria is caused by an enzyme malfunction in the enzyme phenylalanine hydroxylase, which catalyses the first step in the degradation of phenylalanine. If this enzyme does not function, the resulting build-up of phenylalanine leads to mental retardation.

Like all catalysts, enzymes work by lowering the activation energy of a reaction, thus allowing the reaction to proceed much faster. Enzymes may speed up reactions by a factor of many thousands. An enzyme, like any catalyst, remains unaltered by the completed reaction and can therefore continue to function. Because enzymes, like all catalysts, do not affect the relative energy between the products and reagents, they do not affect equilibrium of a reaction. However, the advantage of enzymes compared to most other catalysts is their sterio-, regio- and chemoselectivity and specificity.

Enzyme activity can be affected by other molecules. Inhibitors are molecules that decrease or abolish enzyme activity; activators are molecules that increase the activity. *Suicide inhibitors* are inhibitors that incorporate themselves into the enzyme, permanently deactivating it. Inhibitors can be either natural or man-made. Many drugs are enzyme inhibitors. Aspirin, for example, inhibits an enzyme that produces the inflammation messenger prostaglandin, thus suppıessing pain and inflammation.

Enzymes are also used in everyday products such as washing detergents, where they speed up chemical reactions involved in cleaning the clothes (for example, breaking down starch stains).

More than 5,000 enzymes are known. To name different enzymes, one typically uses the ending *-ase* with the name of the chemical being transformed (substrate), e.g., lactase is the enzyme that catalyzes the cleavage of lactose.

Endergonic and Exergonic

Energy releasing processes, ones that "generate" energy, are termed exergonic reactions. Reactions that require energy to initiate the reaction are known as endergonic reactions. All natural processes tend to proceed in such a direction that the disorder or randomness of the universe increases (the second law of thermodynamics).

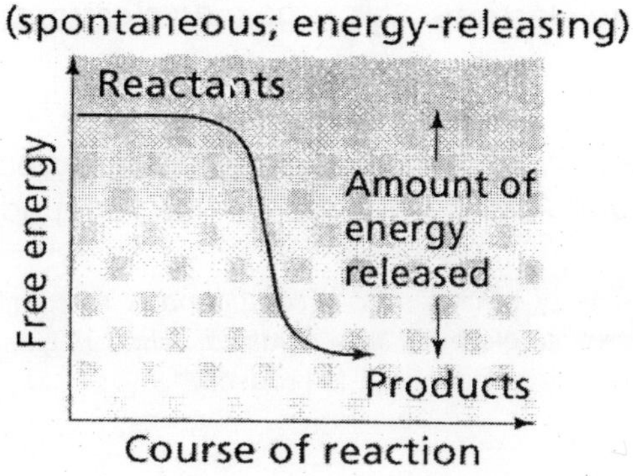

Figure 2: Exergonic reaction

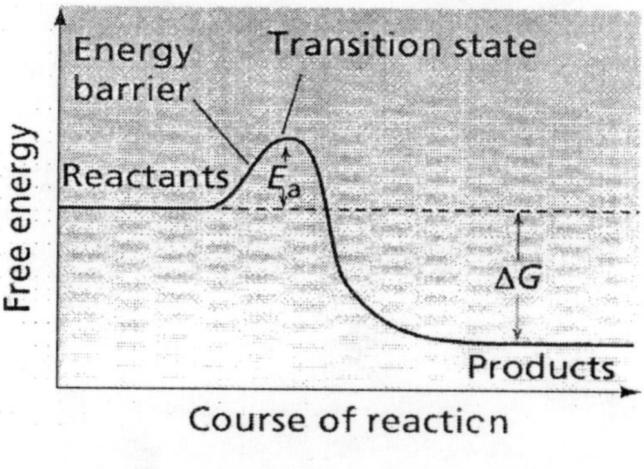

Figure 3: Endergonic reaction

Oxidation/Reduction

Biochemical reactions in living organisms are essentially energy transfers. Often they occur together, "linked", in what are referred to as oxidation/reduction reactions. Reduction is the gain of an electron.

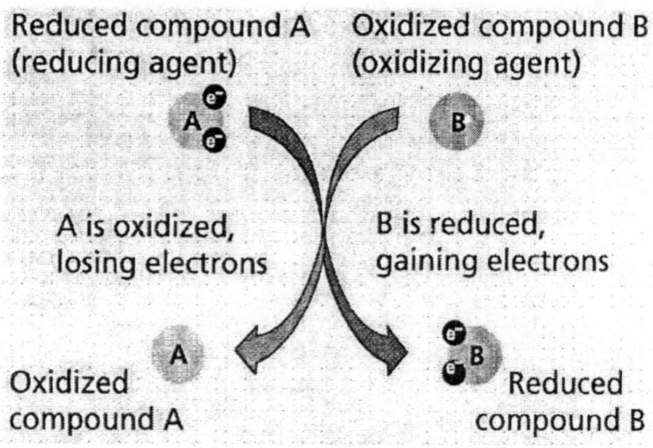

Figure 4: Passage of electrons from compound A to compound B. When A loses its electrons it is oxidized; when B gains the electrons it is reduced.

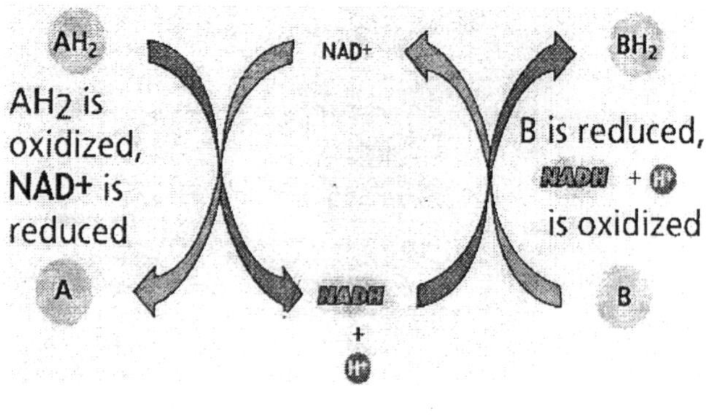

Ctd . . .

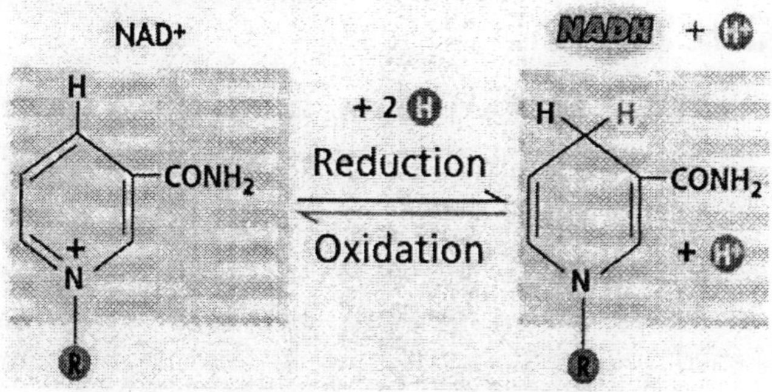

Figure 5: Oxidation/reduction via an intermediary (energy carrier) compound, in this case NAD^+.

Sometimes we also have H ions along for the ride, so reduction also becomes the gain of H. Oxidation is the loss of an electron (or hydrogen). In oxidation/reduction reactions, one chemical is oxidized, and its electrons are passed (like a hot potato) to another (reduced, then) chemical. Such coupled reactions are referred to as redox reactions. The metabolic processes glycolysis, Kreb's Cycle, and Electron Transport Phosphorylation involve the transfer of electrons (at varying energy states) by redox reactions.

Catabolism and Anabolism

Anabolism is the total series of chemical reactions involved in synthesis of organic compounds. Autotrophs must be able to manufacture (synthesize) all the organic compounds they need. Heterotrophs can obtain some of their compounds in their diet (along with their energy). For example humans can synthesize 12 of the 20 amino acids, we must obtain the other 8 in our diet. Catabolism is the series of chemical reactions that breakdown larger molecules. Energy is released this way, some of it can be utilized for anabolism. Products of catabolism can be reassembled by anabolic processes into new anabolic molecules.

Organic Catalysts

Enzymes allow many chemical reactions to occur within the homeostasis constraints of a living system. Enzymes function as organic catalysts. A catalyst is a chemical involved in, but not changed by, a chemical reaction. Many enzymes function by lowering the activation energy of reactions. By bringing the reactants closer together, chemical bonds may be weakened and reactions will proceed faster than without the catalyst.

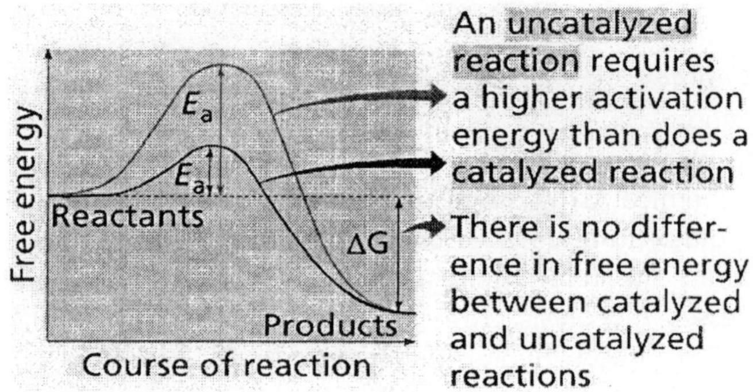

Figure 6: The use of enzymes can lower the activation energy of a reaction (E_a).

Enzymes can act rapidly, as in the case of carbonic anhydrase (enzymes typically end in the -ase suffix), which causes the chemicals to react 107 times faster than without the enzyme present. Carbonic anhydrase speeds up the transfer of carbon dioxide from cells to the blood. There are over 2000 known enzymes, each of which is involved with one specific chemical reaction. Enzymes are substrate specific. The enzyme peptidase (which breaks peptide bonds in proteins) will not work on starch (which is broken down by human-produced amylase in the mouth).

Enzymes are proteins. The functioning of the enzyme is determined by the shape of the protein. The arrangement of molecules on the enzyme produces an area known as the active site within which the specific substrate(s) will "fit". It recognizes, confines and orients the substrate in a particular direction.

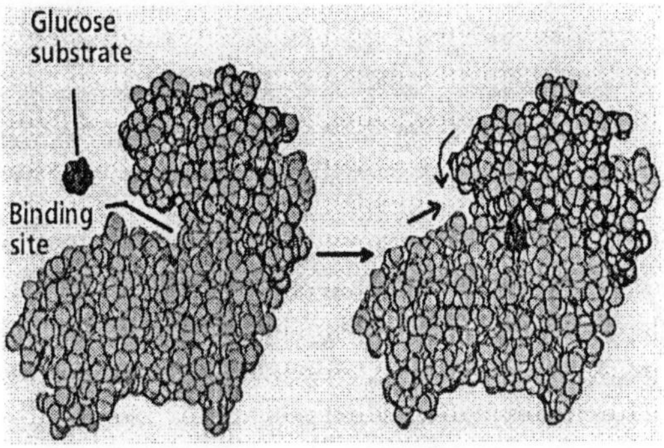

Figure 7: Space filling model of an enzyme working on glucose.

The induced fit hypothesis suggests that the binding of the substrate to the enzyme alters the structure of the enzyme, placing some strain on the substrate and further faci itating the reaction. Cofactors are nonproteins essential for enzyme activity. Ions such as K^+ and Ca^{+2} are cofactors. Coenzymes are nonprotein organic molecules bound to enzymes near the active site.

Enzymatic pathways form as a result of the common occurrence of a series of dependent chemical reactions. In one example, the end product depends on the successful completion of five reactions, each mediated by a specific enzyme. The enzymes in a series can be located adjacent to each other (in an organelle or in the membrane of an organelle), thus speeding the reaction process. Also, intermediate products tend not to accumulate, making the process more efficient.

By removing intermediates (and by inference end products) from the reactive pathway, equilibrium (the tendency of reactions to reverse when concentrations of the products build up to a certain level) effects are minimized, since equilibrium is not attained, and so the reactions will proceed in the "preferred" direction.

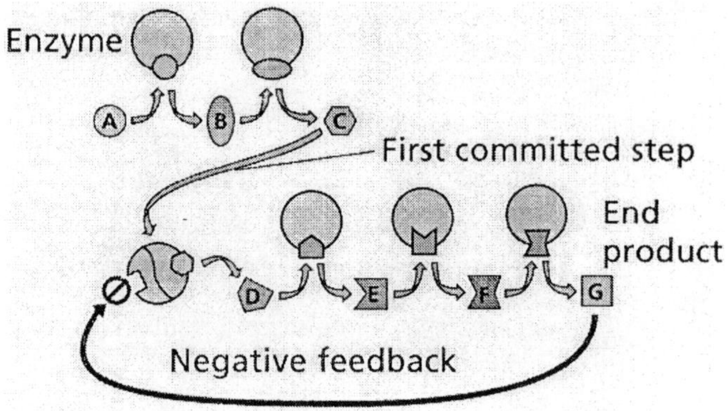

Figure 8: Negative feedback and a metabolic pathway.

The production of the end product (G) in sufficient quantity to fill the square feedback slot in the enzyme will turn off this pathway between step C and D.

Temperature: Increases in temperature will speed up the rate of nonenzyme mediated reactions, and so temperature increase speeds up enzyme mediated reactions, but only to a point. When heated too much, enzymes (since they are proteins dependent on their shape) become denatured. When the temperature drops, the enzyme regains its shape. Thermolabile enzymes, such as those responsible for the color distribution in Siamese cats and color camouflage of the Arctic fox, work better (or work at all) at lower temperatures.

Concentration of substrate and product also control the rate of reaction, providing a biofeedback mechanism. Activation, as in the case of chymotrypsin, protects a cell from the hazards or damage the enzyme might cause. Changes in pH will also denature the enzyme by changing the shape of the enzyme. Enzymes are also adapted to operate at a specific pH or pH range.

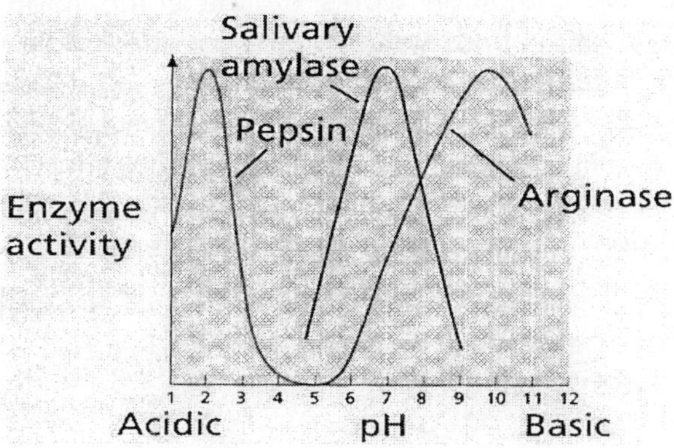

Figure 9: Plot of enzyme activity as a function of pH for several enzymes

Allosteric Interactions may allow an enzyme to be temporarily inactivated. Binding of an allosteric effector changes the shape of the enzyme, inactivating it while the effector is still bound. Such a mechanism is commonly employed in feedback inhibition. Often one of the products, either an end or near-end product act as an allosteric effector, blocking or shunting the pathway.

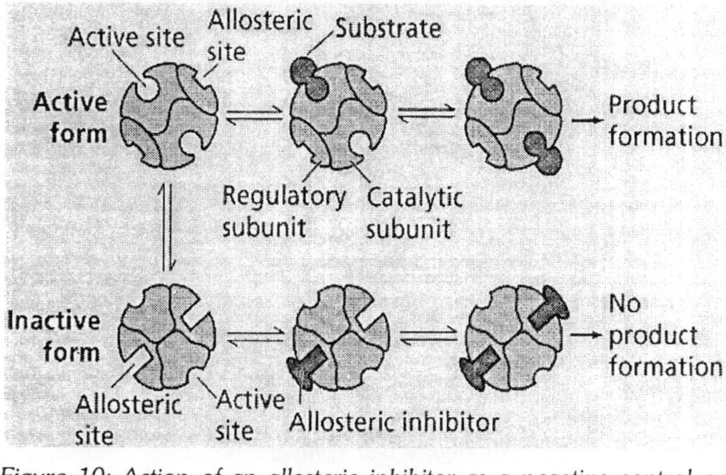

Figure 10: Action of an allosteric inhibitor as a negative control on the action of an enzyme.

Competitive Inhibition works by the competition of the regulatory compound and substrate for the binding site. If enough regulatory compound molecules bind to enough enzymes, the pathway is shut down or at least slowed down. PABA, a chemical essential to a bacteria that infects animals, resembles a drug, sulfanilamide, that competes with PABA, shutting down an essential bacterial (but not animal) pathway.

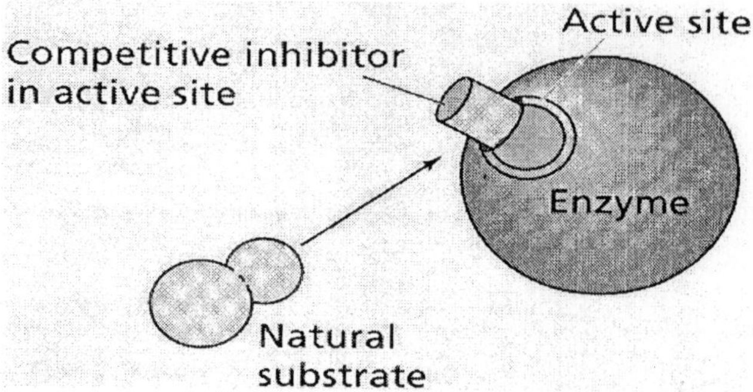

Figure 11: Competitor in the active site normally occupied by the natural substrate

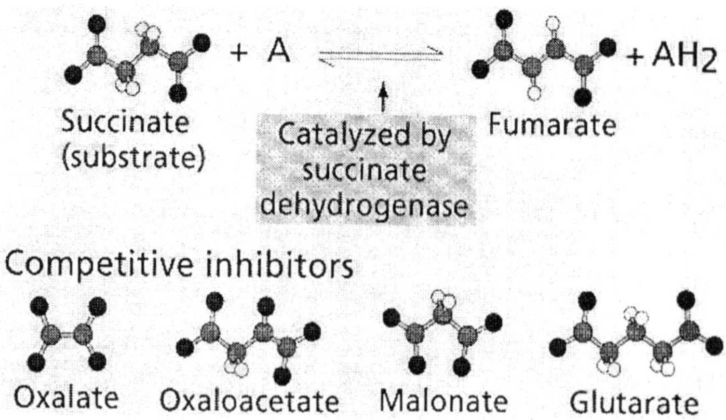

Figure 12: Succinate dehydrogenase and its natural substrate (succinate) and competitors

Noncompetitive Inhibition occurs when the inhibitory chemical, which does not have to resemble the substrate, binds to the enzyme other than at the active site. Lead binds to SH groups in this fashion. Irreversible Inhibition occurs when the chemical either permanently binds to or massively denatures the enzyme so that the tertiary structure cannot be restored. Nerve gas permanently blocks pathways involved in nerve message transmission, resulting in death. Penicillin, the first of the "wonder drug" antibiotics, permanently blocks the pathways certain bacteria use to assemble their cell wall components.

Catalysts in Reactions

Catalysts participate in reactions but are not consumed. They increase reaction rates by lowering activation energy, E_A. The key is that the reactants must be adsorbed briefly onto the surface or a crack in the catalyst. So, catalysis is a surface-dependent phenomenon. The adsorption or binding between reactants and catalyst does not take place in a haphazard manner. Instead, what makes the surface of a catalyst effective is that it binds the reactants in the correct steric (position) relationships to allow the reaction to occur.

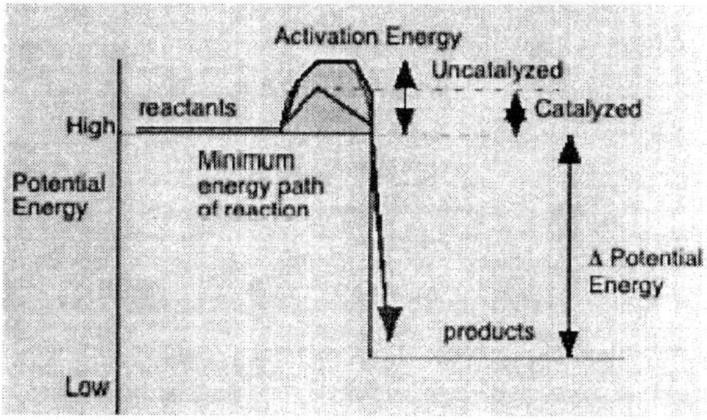

Figure 13: Energy hill diagram

The binding of the reactants to the catalyst is typically via some sort of polar or covalent interaction-or, sometimes it is the result of a hydrophobic interaction. For the moment, let's assume the reactant(s) bind initially via polar interaction (the most common case). The binding often creates a stress in the bond(s) where the reaction is to take place. You can think of the reactant molecule as being distorted somewhat by the catalyst. This is the equivalent of what happens when we add lots of activation energy-recall from the last notes that the increased energy caused, randomly, much more stress and distortion in the bonds of the reactants.

If there is more than one reactant, the other thing the catalyst does is to ensure that the reactants are very close to each other and in the correct position for the specific reaction that is being catalyzed. The catalyst has done-at a much lower system energy, it has stressed the bonds in the reactants and brought them together in the correct configuration to encourage a specific reaction-simply by only allowing them to bind in a certain configuration. This has happened with a much lower requirement of activation energy.

The reaction was very specific-there was an exact mechanism leading to an exact product. To illustrate how specific catalyzed reactions usually are-recall that in many reactions involving organic molecules, it may be possible to produce two optical isomers of what is otherwise the same compound. In uncatalyzed reactions, both will be produced. However, catalysts generally only produce one. After a reactant binds loosely to the catalyst (for example, by hydrogen bonds or via a hydrophobic interaction, keep in mind the fact that sometimes the reactant will briefly become covalently bonded to the catalyst. This is quite common in enzymes. This covalent binding is generally part of the reaction mechanism.

A reactant becomes covalently bonded, its characteristics change because of this bonding, and then it reacts with the other reactant. In the process, the covalent bond between the initial reactant and the catalyst break. Once the reaction is complete, the products are released. The reactants have been converted to products at a much lower energy level than would be possible than without the catalyst. Also, the catalyst has not been used up (unlike the reactants). It is there and now can be used again.

On the macro level that a thermodynamically feasible reaction is proceeding under physical conditions (most likely low temperature or pressure) where it normally would not. In other words, the catalyst has altered the reaction kinetics by lowering E_A. Again, the reaction is occurring at energy levels that are too low to support a noticeable reaction in the absence of the catalyst. Finally, note that the reaction rate will be proportional to the amount of reactant present, the "distance" from equilibrium, and most importantly, the amount of catalytic binding sites.

Enzymes-Protein Catalysts

Proteins can come in a bewildering variety of shapes. Their higher order structures are influenced by interactions between their amino acids with each other and with the broader physical and chemical environment. When a polypeptide gains tertiary structure, it is not uncommon for it to have clefts or openings leading to cavities of various shapes. In principle, any shape might be possible. Beyond shape, the amino acids that line the inside of any such cavities can have many different characteristics. Some parts of the cavity wall could be polar, others non-polar. There could be regions where the amino acid side groups are relatively reactive with certain other compounds, other regions where they are inert.

Binding

The first step is for the enzyme and substrate to bind. Initially, this binding is weak and is accomplished by complementary shapes of enzyme and substrate, hydrogen bonding in most cases, and in some cases by regions of hydrophobic interaction. Due to the weakness of these interactions, the enzyme-substrate complex is relatively easily broken up-it is easily reversible. We can view the binding step as:

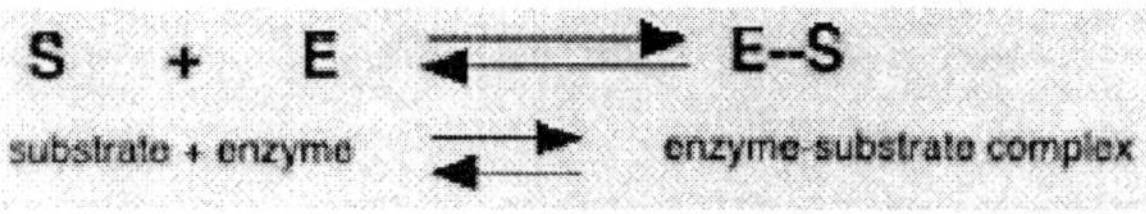

Binding occurs as a result of collisions between substrate and active site. Not all collisions will result in an enzymesubstrate complex. That is because the substrate(s) may not hit the active site in the correct (or nearly correct configuration. Moreover, there will be many cases when some factor will have caused the shape of the active site to be altered in a way that it is not a particular good fit for the substrate (in fact, initially that is always the case). In addition, once the E-S complex has been formed, there is no guarantee that it will persist long enough for a reaction to occur.

If there is not a particularly good shape fit between enzyme and substrate (for example, say some environmental factor such as a change in pH has altered the tertiary structure of the enzyme somewhat), the substrate could easily be knocked back out again. One other important thing happens during binding of substrate and enzyme. The shape of both the enzyme and the now bound substrate molecules change somewhat. This is very important.

First, it is exactly what we would expect about a tertiary structure such as an enzyme. These are huge molecules where much of the shape is owed to relatively weak interactions. The addition and binding of (an)other molecule(s) to the enzyme (substrate(s)) will have to affect the overall shape of the enzyme. Likewise, a shape change in the large enzyme molecule will surely affect the shape of the small regions of substrate bound within the active site.

The usual result of this initial shape change is to increase the strength of binding between enzyme and substrate align various side groups within the active site with the substrate in a way that will promote the reaction about to be catalyzed, and realign the substrate molecules with respect to each other. Moreover, the bending of the substrates tends to stress certain bonds and make them more reactive.

We call this sort of an adaptive shape change in a protein an ALLOSTERIC CHANGE. Initially E and S do not bind well, but with interaction, an allosteric change results in better binding and the potential for rapid catalysis. We will use the creatine kinase reaction as an example. Creatine kinase is an enzyme that catalyzes the following reaction:

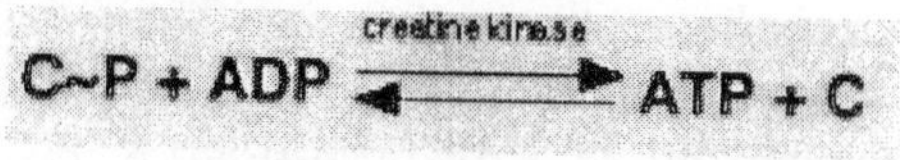

where C~P is creatine phosphate (essentially a molecule that stores high-energy phosphates for times of high-energy demand), C is creatine, and you already know ADP and ATP.

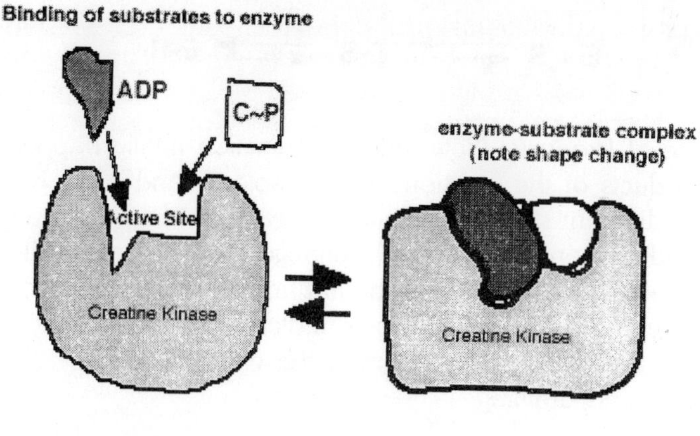

Figure 14: Depiction of the binding step

During periods where cells (muscle in this case) demand ATP at a higher rate than it can be produced, ATP decreases and

ADP increases. However, ATP is what various contractile proteins in the muscles require. To make up for the shortfall in ATP, cells transfer a high-energy phosphate from C~P to ADP to give ATP. Its kind of like taking money out of your bank account when you spend faster than you earn (where ATP represents a filled wallet and C~P is the banked money).

Catalysis

Due to the allosteric change, the substrate molecule(s) is (are) now in a position to react. In many cases, the actual reaction is a bit complicated-for instance, one substrate may actually react and produce a covalent bond with one of the side groups of the amino acids lining the active site. Then, part or all of the molecule is again transferred to complete the reaction. However, any covalent bonds between enzyme and substrate(s) are broken by the time the reaction is complete. The enzyme is now, chemically speaking, back to normal. We can talk of there being an enzyme-product (E-P) complex. So, here is the sequence so far:

$$E + S \rightleftharpoons E\text{-}S \rightleftharpoons E\text{-}P$$

Release of Products: Now, after the reaction finally happens, the products of the reaction (for example, C and ATP) have slightly different shapes than did the reactants a moment ago. This induces another allosteric change in the enzyme. The change lowers the affinity of the enzyme for the products and they are released. The enzyme molecule is now ready to catalyze another reaction. Here is the entire sequence as an equation and drawing:

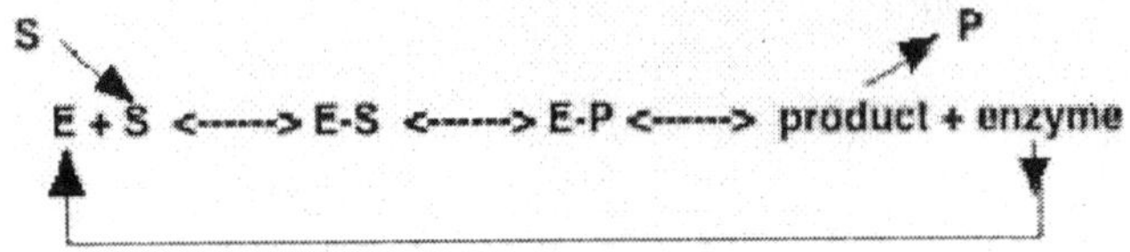

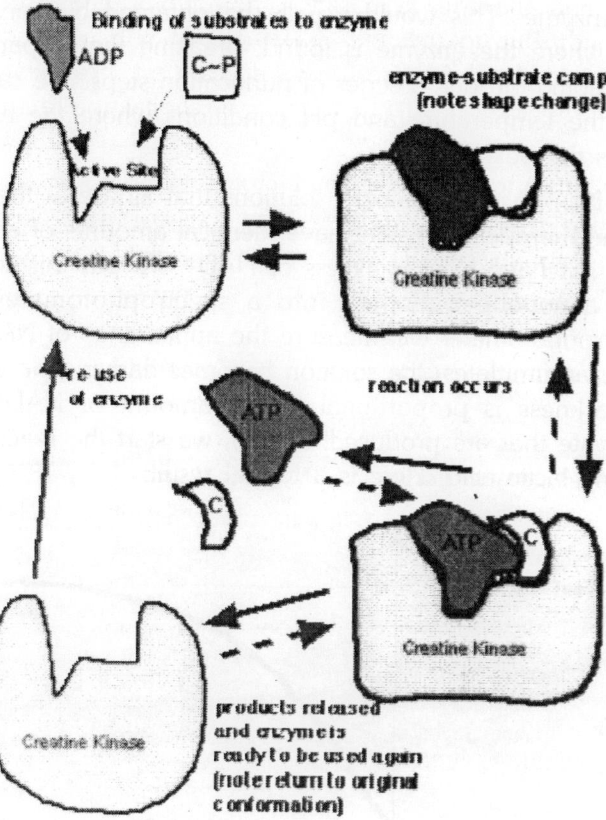

Kinetics of Enzymatic Reactions

Let's suppose that we want to investigate the properties of the enzymes. The specific enzyme lactic dehydrogenase (LDH) is the experimental subject. It catalyzes the following reaction:

$$\text{lactate} + NAD^+ \underset{}{\overset{LDH}{\rightleftharpoons}} \text{pyruvate} + NADH + H^+$$

To do the investigation we can obtain a sample that contains the enzyme. This would usually be obtained by harvesting cells where the enzyme is found, breaking them apart, and then going through a series of purification steps. We can also find the temperature and pH conditions where the enzyme works the best.

Next, we can prepare solutions that all are at the same pH and temperature, that have identical amounts of enzyme, and that have a large excess of NAD^+. We put this solution into a cuvette and then into a spectrophotometer. The spectrophotometer will measure the appearance of NADH-as more accumulates, the solution becomes darker. The change in darkness is proportional to the amount of NADH and pyruvate that are produced. Finally, we start the reaction by adding lactic acid. Here is a typical result:

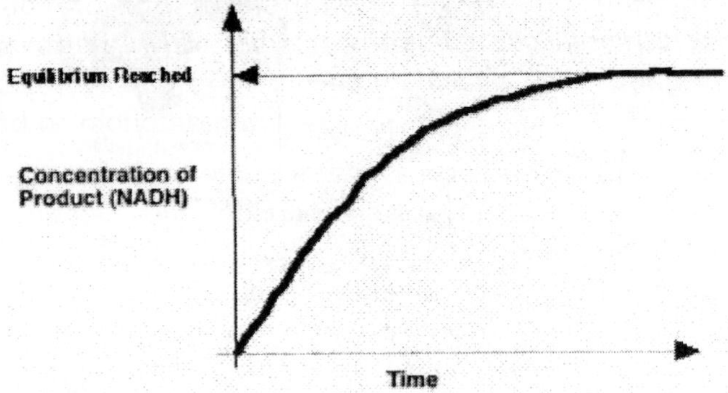

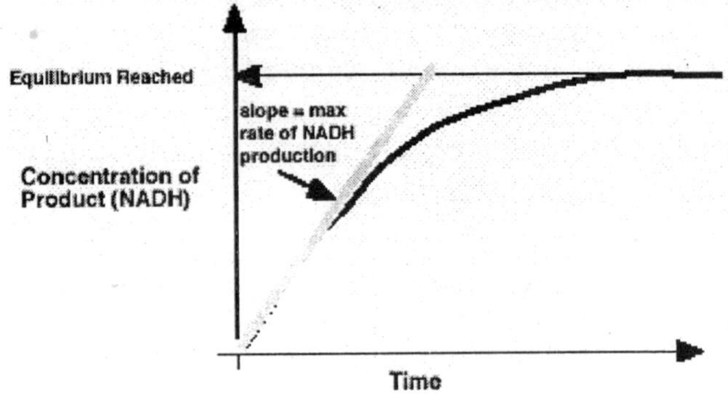

Now, repeat the measurement under identical conditions. Here is an example of the results of the previous measurements combined with another set of measurements at a lower substrate (lactate) concentration. The concentration of enzyme is the same for both:

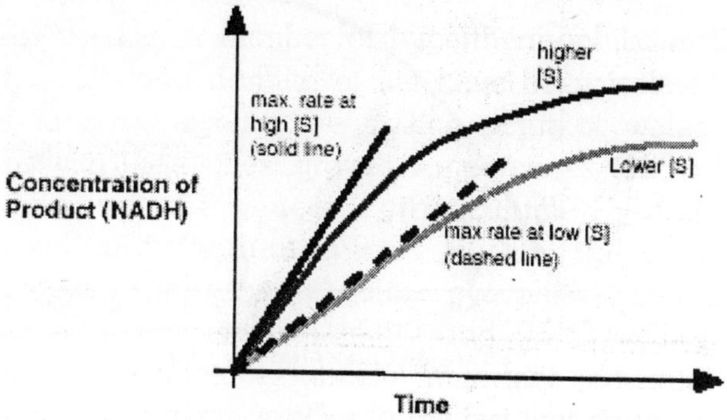

Notice that the maximum reaction rate (given by the slope of the tangent line for each curve) is less when there is less substrate present. On the other hand, if we continue increasing the amount of lactic acid, we will reach a point where additional lactate no longer causes a greater maximum reaction rate:

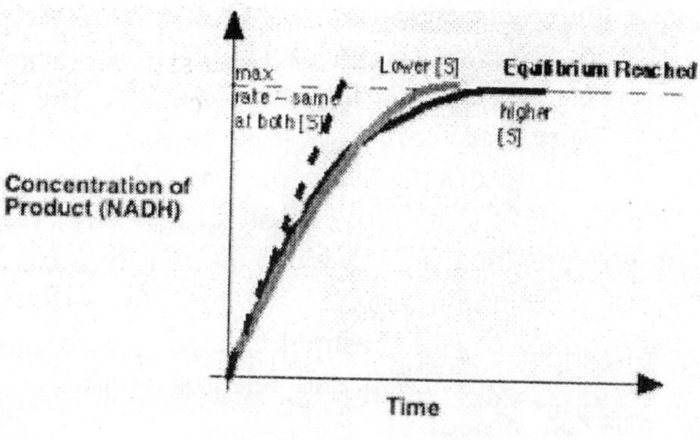

Notice that the maximum rate in these cases is much greater than in the previous ones. Now, we take the maximum rate for each [lactate] and make a plot:

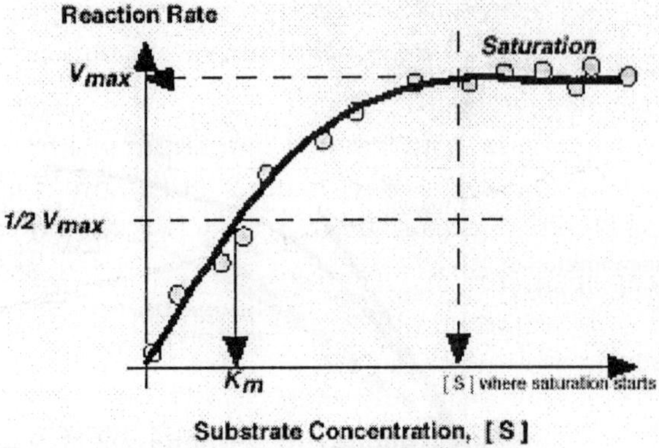

Notice what this graph shows us. For a constant amount of enzyme:

— Up to a point the reaction rate increases with an increase in substrate

— Accordingly, there also exists a [substrate] which when exceeded does not result in any further increase in the reaction rate (for a given amount of enzyme)

At very low, the reaction rate is low because there is so little substrate. Under these conditions, what substrate there is will have no trouble finding an enzyme molecule. The rate is low simply because there is little substrate present, not because it is hard to find an enzyme molecule. As [S] increases, the reaction rate increases since more substrate molecules per time are finding enzymes that then catalyze the reaction. However, if the [S] is sufficiently high, every enzyme molecule is continuously occupied with substrate catalyzing a reaction. As fast as it completes a reaction and releases products, a new substrate molecule arrives.

It is like a production line worker-it takes a certain amount of time to do the job it is supposed to do. When the conveyer belt moves slowly the worker is not always working and the output of the production line is low. As the belt speeds up, the worker spends more time occupied and the output increases. However, a point is reached where the line goes so fast that the worker cannot work any faster. S/he will let some items pass without working on them. The output of the line will not decrease but nor will it increase. After reaching saturation, adding more substrate (the equivalent of speeding up the line) will not result in any increase in reaction rate.

Wokring backward from the saturation point we can arbitrarily define another point which will have great usefulness to us. This point, called the Michelis constant (K_M) is the [substrate] at ½ V_{MAX}. We will use the K_M as a measure of the affinity of the enzyme for this substrate. The more enzyme molecules of a certain type are present, then, for a given concentration of substrate, the faster the reaction will go.

Therefore, a good way to increase reaction rate is to add more enzyme. Now, what if we use different versions of the same enzyme-i.e., enzymes that catalyze the same reaction but that have slightly different tertiary structures. Such situations are common. There are two ways to get differences in enzyme tertiary structure:

— Different primary structures-these are normally called allozymes or isozymes; they are produced by different alleles for the same gene; i.e., by slightly different versions of the same gene. We have already seen that differences in primary structure, when expressed in the same cellular environment, result in differences in tertiary structure. For the case of lactic dehydrogenase, your body produces several different versions of the enzyme that are based on slight differences in primary structure. You have one kind in your skeletal muscles and another slightly different form in your heart.

— The same primary structure but different environments-we could alter the chemical and/or physical environment in which the enzyme was formed or is being maintained. The result would be differences in tertiary structure.

Chapter 2

Enzyme Inhibition

Since structure mediates function, anything that would significantly change the structure of an enzyme would inhibit the activity of the enzyme. Hence extremes of pH and high temperature, all of which can denature the enzyme, would inhibit the enzyme in a irreversible fashion, unless it could refold properly. Alternatively we could add a small molecule which interacts noncovalently with the enzyme to either change its conformation or directly prevent substrate binding.

The types of reagents that would chemically modify specific side chains that might be critical for enzymatic activity. For example, iodoacetamide might abolish enzyme activity if a Cys side chain is required for activity. These reagents will usually modify several side chains, however, and determining which is critical for binding or catalytic conversion of the substrate can be difficult.

One way would be to protect the active site with a saturating quantities of a ligand which binds reversibly at the active site. Then the chemical modification can be performed at varying reaction times. The critical side chain would be protected from the chemical modification, but the extent of protection would depend on the Kd, concentration of the protecting ligand, and the length of the reaction.

Competitive Inhibition

Competitive inhibition occurs when substrate (S) and inhibitor (I) both bind to the same site on the enzyme. In effect, they compete for the active site and bind in a mutually exclusive fashion. This is illustrated in the mechanism at the top of the figure to the left, and in the molecular cartoon beneath it.

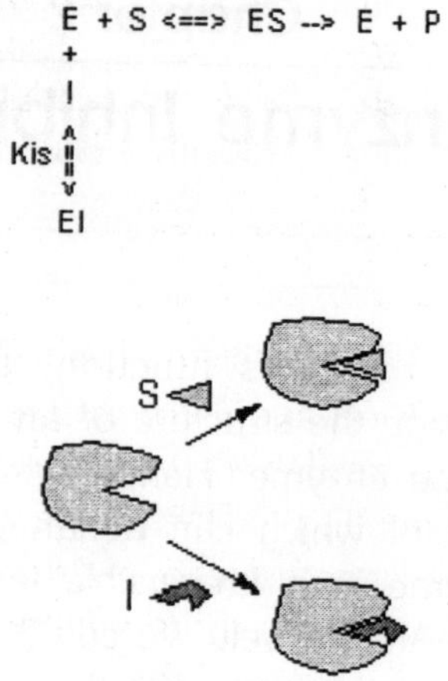

$$v = \frac{VmS}{Km(1 + I/Kis) + S}$$

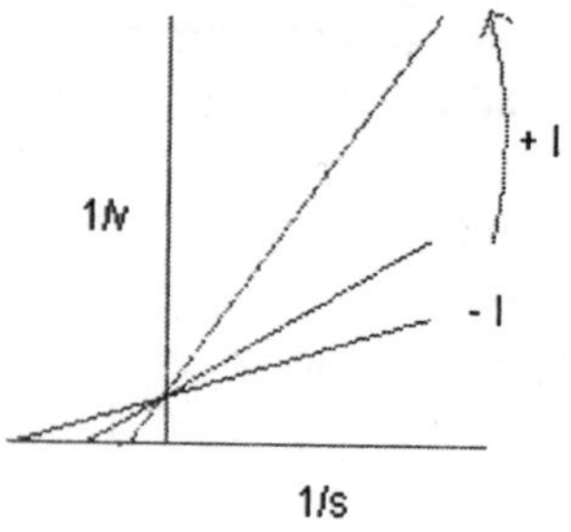

Figure 1: Competive inhibition

There is another type of inhibition that would give the same kinetic data. If S and I bound to different sites, and S bound to E and produced a conformational change in E such that I could not bind (and vice versa), then the binding of S and I would be mutually exclusive. This is called allosteric competitive inhibition. Inhibition studies are usually done at several fixed and non-saturating concentrations of I and varying S concentrations.

The key kinetic parameters to understand are Vm and Km. Let us assume for ease of equation derivation that I binds reversibly, and with rapid equilibrium to E, with a dissociation constant K_{is}. A look at the top mechanism shows that even in the presence of I, as S increases to infinity, all E is converted to ES. That is, there is no free E to which I could bind. Now remember that Vm= k_{cat}Eo. Under these condition, ES = E_o; hence v = Vm.

Vm is not changed. However, the apparent Km, Km_{app}, will change. We can use LaChatelier's principle to understand this. If I binds to E alone, and not ES, it will shift the equilibrium of E + S <==> ES to the left, which would have the affect of increasing the Km_{app} (i.e. it would appear that the affinity of E and S has decreased.). The double reciprocal plot (Lineweaver Burk plot) offers a great way to visualize the inhibition. In the presence of I, Vm does not change, but Km appears to increase. Therefore, 1/Km, the x-intercept on the plot will get smaller, and closer to 0. Therefore the plots will consists of a series of lines, with the same y intercept (1/Vm), and the x intecepts (-1/Km) closer and closer to the 0 as I increases. These intersecting plots are the hallmark of competitive inhibition.

An equation, shown in the diagram above, can be derived which shows the effect of the competitive inhibitor on the velocity of the reaction. The only change is that the Km term is multiplied by the factor $1+I/K_{is}$. Hence Kmapp = $Km(1+I/K_{is})$. This shows that the apparent Km does increase as we predicted. K_{is} is the inhibitor dissociation constant in which the inhibitor affects the **s**lope of the double reciprocal plot.

Uncombetitive Inhibition

Uncompetitive inhibition occurs when I binds only to ES and not free E. One can hypothesize that on binding S, a conformational change in E occurs which presents a binding site for I. Inhibition occurs since ESI can not form product. It is a dead end complex which has only one fate, to return to ES.

$$E + S \Longleftrightarrow ES \longrightarrow E + P$$

$$ES + I \overset{K_{ii}}{\Longleftrightarrow} ESI \not\longrightarrow E + P$$

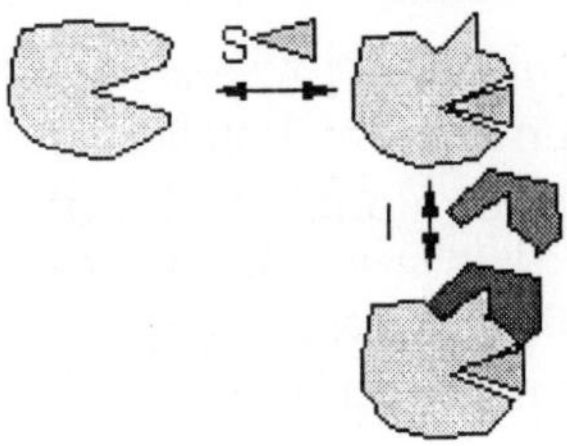

$$v = \frac{VmS}{Km + S(1 + I/Ki)}$$

$$v = \frac{\left(\frac{Vm}{1+I/Kii}\right) S}{\left(\frac{Km}{1+I/Kii}\right) + S}$$

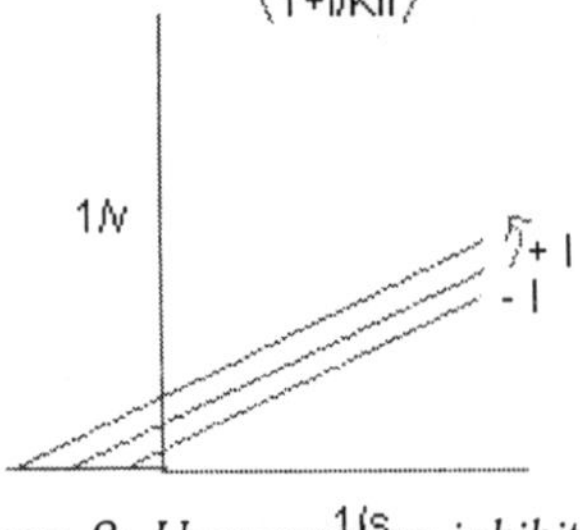

Figure 2: Uncompetitive inhibition

This is illustrated in the mechanism at the top of the figure to the left, and in the molecular cartoon beneath it. Let us assume for ease of equation derivation that I binds reversibly to ES with a dissociation constant K_{ii}. A look at the top mechanism shows that in the presence of I, as S increases to infinity, not all of E is converted to ES. That is, there is a finite amount of ESI, even at infinite S. Now remember that $Vm = k_{cat}E_o$ if and only if all E is in the form ES . Under these conditions, the apparent Vm, Vm_{app} is less than the real Vm without inhibitor.

In addition, the apparent Km, Km_{app}, will change. We can use LaChatelier's principle to understand this. If I binds to ES alone, and not E, it will shift the equilibrium of E + S <==> ES to the right , which would have the affect of decreasing the Km_{app} (i.e. it would appear that the affinity of E and S has increased.). The double reciprocal plot (Lineweaver Burk plot) offers a great way to visualize the inhibition. In the presence of I, both Vm and Km decrease. Therefore, -1/Km, the x-intercept on the plot, will get more negative, and 1/Vm will get more positive. It turns out that they change to the same extent. Therefore the plots will consist of a series of parallel lines, which is the hallmark of uncompetitive inhibition.

An equation, shown in the diagram above, can be derived which shows the effect of the uncompetitive inhibitor on the velocity of the reaction. The only change is that the S term in the denominator is multiplied by the factor $1+I/K_{ii}$. We would like to rearrange this equation to show how Km and Vm are affected by the inhibitor, not S, which obviously isn't. Rearranging the equation as shown above shows that $Km_{app} = Km/(1+I/K_{ii})$ and $Vm_{app} = Vm/(1+I/K_{ii})$. This shows that the apparent Km and Vm do decrease as we predicted. K_{ii} is the inhibitor dissociation constant in which the inhibitor affects the intercept of the double reciprocal plot. Note that if I is zero, Km and Vm are unchanged.

Noncompetitive Inhibition

Noncompetitive inhibition occurs when I binds to both E and ES. The special case in which the dissociation constants of I for E and ES are the same. This is called noncompetetiive inhibition. In the more general case, the Kd's are different, and the inhibition is called mixed.

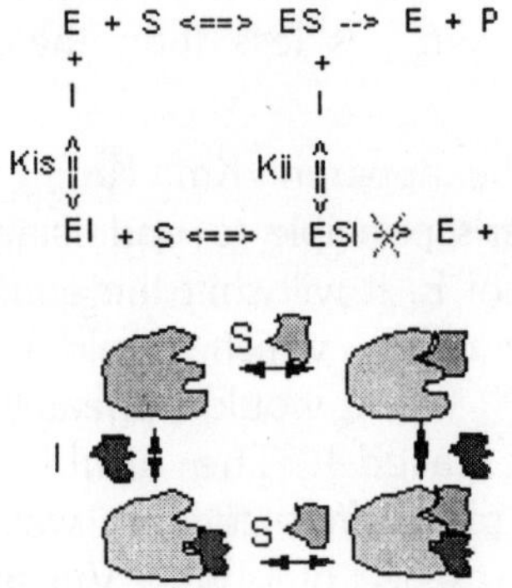

$$v = \frac{VmS}{Km(1+I/Kis) + S(1+I/Kii)}$$

$$v = \frac{\left(\frac{Vm}{1+I/Kii}\right)S}{Km + S} \quad \text{if } Kii = Kis$$

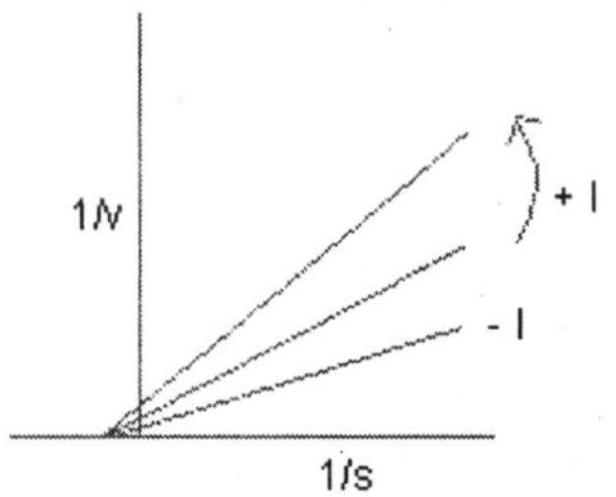

Figure 3: Non-competive inhibition

Since inhibition occurs, we will hypothesize that ESI can not form product. It is a dead end complex which has only one fate, to return to ES or EI. This is illustrated in the mechanism at the top of the figure to the left, and in the molecular cartoon beneath it. Let us assume for ease of equation derivation that I binds reversibly to E with a dissociation constant of K_{is} and to ES with a dissociation constant K_{ii}.

Assume for noncompetitive inhibition that $K_{is} = K_{ii}$. A look at the top mechanism shows that in the presence of I, as S increases to infinity, not all of E is converted to ES. That is, there is a finite amount of ESI, even at infinite S. Now remember that $Vm = k_{cat}E_o$ if and only if all E is in the form ES . Under these conditions, the apparent Vm, Vm_{app} is less than the real Vm without inhibitor. In contrast, the apparent Km, Km_{app}, will not change since I binds to both E and ES with the same affinity, and hence will not perturb that equilibrium, as deduced from LaChatelier's principle.

The double reciprocal plot (Lineweaver Burk plot) offers a great way to visualize the inhibition. In the presence of I, just Vm will decrease. Therefore, -1/Km, the x-intercept will stay the same, and 1/Vm will get more positive. Therefore the plots will consists of a series of lines intersecting on the x axis, which is the hallmark of noncompetitive inhibition. You should be able to figure out how the plots would appear if K_{is} is different from K_{ii} (mixed inhibition).

An equation, shown in the figure above can be derived which shows the effect of the uncompetitive inhibitor on the velocity of the reaction. In the denominator, Km is multiplied by $1+I/K_{is}$, and S by $1+I/K_{ii}$. To rearrange this equation to show how Km and Vm are affected by the inhibitor, not S, which obviously isn't. Rearranging the equation as shown above shows that $Km_{app} = Km(1+I/K_{is})/(1+I/K_{ii}) = Km$ when $K_{is}=K_{ii}$, and $Vm_{app} = Vm/(1+I/K_{ii})$. This shows that the Km is unchanged and Vm decreases as we predicted. The plot shows a series of lines intersecting on the x axis. Both the slope and the y intercept are changed, which are reflected in the names of the two dissociation constants, K_{is} and K_{ii}.

Temperature Change

From 0 to about 40-50°C, enzyme activity usually increases, as do the rates of most reactions in the absence of catalysts. At higher temperatures, the activity decreases dramatically as the enzyme denatures.

Temperature and enzyme activity

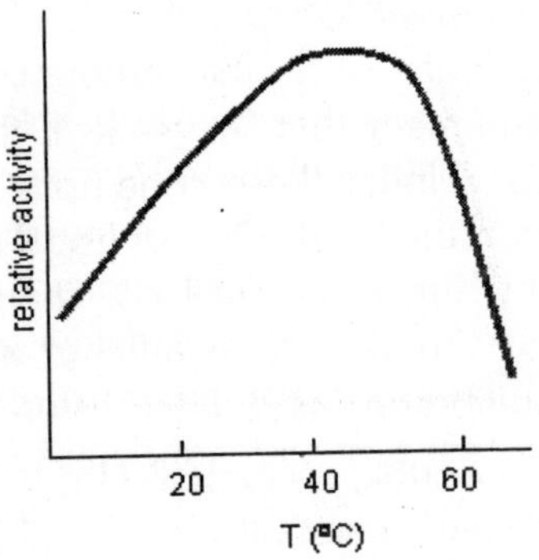

Figure 4: Temperature and Enzyme Activity

Inhibirion by PH Change

pH has a marked effect on the velocity of enzyme-catalyzed reactions.

Think of all the things that pH changes might affect. It might

— affect E in ways to alter the binding of S to E, which would affect Km

— affect E in ways to alter the actual catalysis of bound S, which would affect kcat

— affect E by globally changing the conformation of the protein

— affect S by altering the protonation state of the substrate

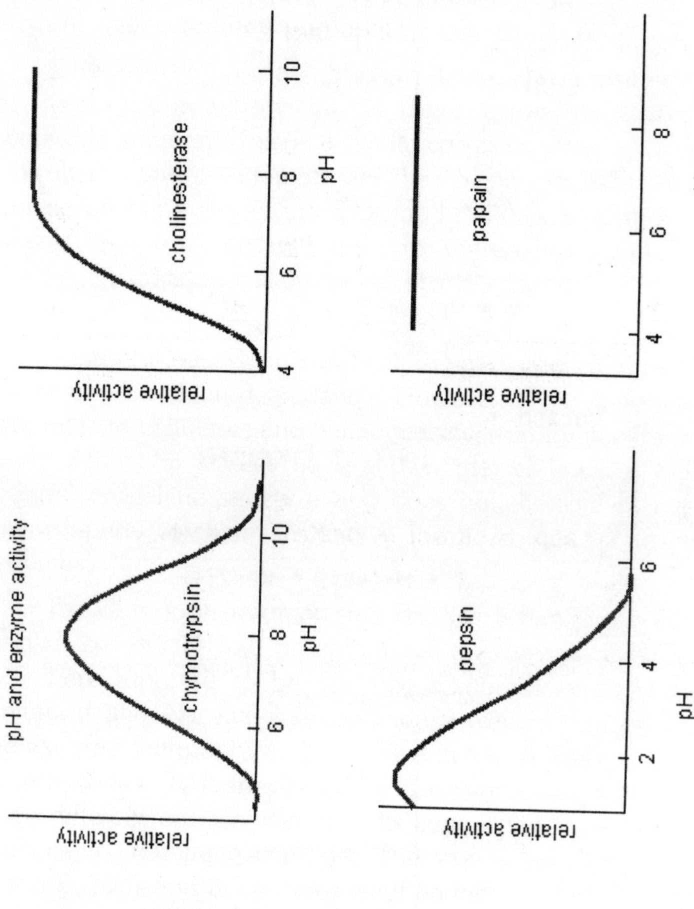

Figure 5: pH and Enzyme Activity

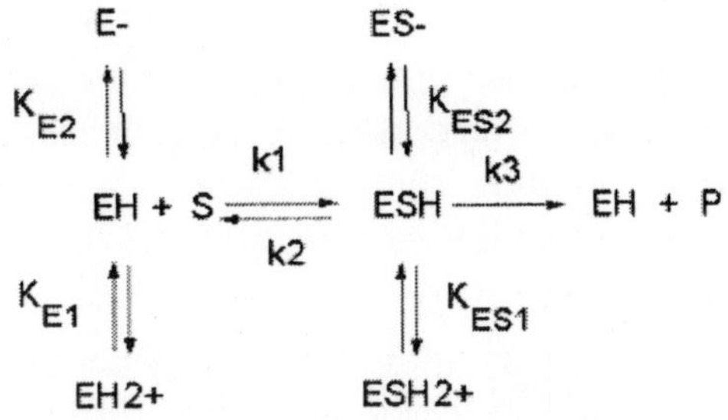

Figure 6: Chemical equations showing the mechanism of pH effects on enzyme catalyzed reactions

$$v = \frac{Vm\ app\ S}{Km\ app + S}$$

$$Vm\ app = \frac{Vm}{1 + H+/Kes1 + Kes2/H+}$$

$$Km\ app = \frac{Km\ (1 + H+/Ke1 + Ke2/H+)}{1 + H+/Kes1 + Kes2/H+}$$

Figure 7: Mathematic equations modeling pH effects on enzyme catalyzed reactions

P

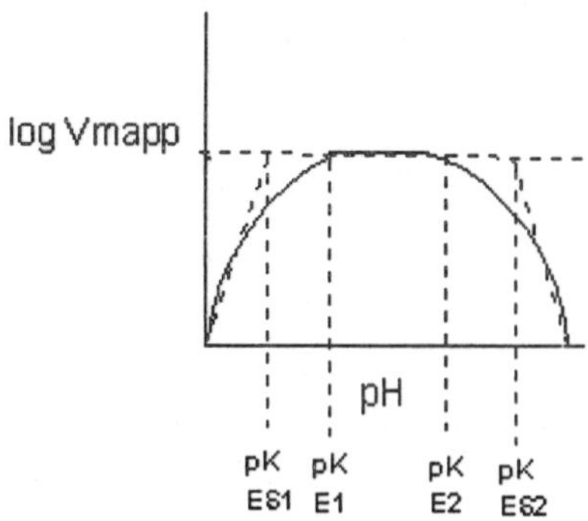

Ctd . . .

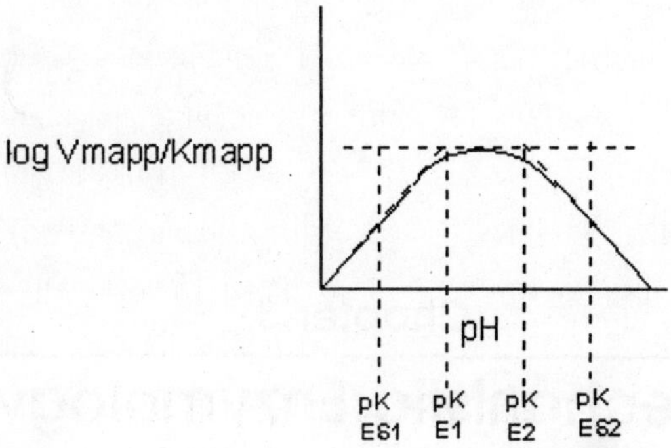

Figure 8: Graphs of pH effects on enzyme catalyzed reactions

The easiest assumption is that certain side chains necessary for catalysis must be in the correct protonation state. Thus, some side chain, with an apparent pKa of around 6, must be deprotonated for optimal activity of trypsin which shows an increase in activity with the increase centered at pH 6. The above figure which shows how pH effects on enzyme kinetics can be modeled at the chemical and mathematical level.

Chapter 3

Mechanistic Enzymology

Lyophilized enzyme powders, suspended in anhydrous organic solvents, have already been used to catalyze ester synthesis, transesterification, peptide synthesis, racemate resolution, epoxidation, hydroxylation, dehydrogenation, polymerization, and depolymerization reactions. Before embarking on any kinetic investigations of enzymes in organic solvents it is necessary to understand any effect of particle morphology (one must always remember that non-aqueous biocatalysis is generally heterogenous) and protein structure on the kinetic behavior of the protein. Suffice it to say here that proteins suspended in organic solvents do not undergo drastic morphological changes as a result of the presence of non-aqueous media.

The particle shape and porosity depends to some degree on the solvent, although most protein powders appear to be either flaky or spherical particles with dimensions of around 100 um in diameter. The particles appear to be both macro- and microporous. Importantly, the addition of water to dry protein powders results in agglomeration and swelling of the particles, combined with a decrease in the microporosity as viewed by environmental electron microscopy.

Interpretation of electron micrographs enables one to determine what types of diffusion must be considered when dealing with heterogenous biocatalysis. Diffusional limitations on heterogenous reactions have been considered for many years, and are the subject of many excellent texts. There are two major diffusional constraints in the non-aqueous enzyme systems which are used most frequently . This is particularly important when attempting to derive correlations between solvent properties and enzyme activity and specificity over a large range of enzyme activities. One must insure that as the enzyme becomes more active, there is no change in the diffusional limitations on the reaction.

Another reason that diffusional limitations must be elucidated before kinetic experiments is that attempts to optimize activity by either solvent or protein engineering are futile if the reaction is under diffusional limitation. Thus, optimization of enzyme function in anhydrous environments must address mass transfer constraints fully. Unfortunately, for those who work in the field there can be no short cuts.

A demonstration that one enzyme in a given environment is not diffusionally limited does not infer that the same enzyme will remain kinetically controlled in another environment. Each powdered enzyme preparation must be tested for external and internal diffusional limitations in each different environment.

A typical comment of classical enzymologists when considering enzyme-catalyzed reactions in non-aqueous media is that the enzyme suspended in the solvent must have a different structure than the same enzyme in water. Indeed, it has been suggested on many occasions that a structural difference accounts for changes in the activity of the enzyme when transferred to non-aqueous environments. Experimentation has shown, however, that any structural changes are minimal.

FT-IR has been used to determine the secondary structure content of subtilisin and myoglobin in a variety of environments. There is no discernable change between subtilisin solubilized in buffer, lyophilized subtilisin, lyophilized subtilisin suspended in mineral oil, and lyophilized subtilisin suspended in carbon tetrachloride. The same held true for myoglobin. Importantly, equivalent experiments on enzyme solubilized in dimethylsulfoxide demonstrate that organic solvents that solvate proteins denature the protein considerably.

The type of denaturation referred to here is markedly different than denaturation by boiling the protein. Finally, the local structure of the myoglobin active site has only minor changes in structure when the protein is placed in anhydrous environments. These experiments were performed by using FT-IR on azidometmyoglobin. Given that enzymes function in organic solvent without a drastic change in structure and that there are significant advantages for such uses, there are still two drawbacks for kinetic experimentation. The most significant is that in most cases the K_m of the enzyme for most substrates is increased upon transfer to the solvent.

Most hydrophobic substrates, which are those with which non-aqueous enzymology would be advantageous, bind to enzyme active sites via hydrophobic bonding. When the bulk solvent is not water, the entropy gain of forming a hydrophobic bond will be significantly decreased and thus Km is expected to increase. For commercial biocatalysis this represents an advantage since high substrate concentrations will not lead to enzyme inactivation. Detailed kinetics cannot be performed unless the enzyme can be saturated with substrate.

The solubility of substrates is often reached before saturation, and in addition if it is possible to saturate the enzyme the use of high concentrations of substrate may alter the physical properties of the solvent relative to pure solvent. Such effects have largely been overlooked in the literature. Because of these limitations it is often impossible to separate the turnover number and Michaelis constant from the specificity constant. Thus, model enzymes which can approach saturption with a number of substrates, and then attempt to understand how the enzyme environment can be modified to predictably alter enzyme activity, specificity and stability.

Subtilisin, a serine protease, is an ideal model enzyme. In water subtilisin catalyzes hydrolysis via the acyl-enzyme mechanism, whereas in organic solvent alcoholysis reactions can be catalyzed. The mechanism is well understood in aqueous solutions, the structure of the enzyme has been determined, and subtilisin is readily available. Furthermore, it is possible to perform site-directed mutagenesis on subtilisin since the gene encoding the protein has been cloned previously. The mechanism of the enzyme in organic solvents has been investigated before.

During the acyl-enzyme mechanism a covalent bond is formed between the serine-221 hydroxyl group of the enzyme, and the substrate. Protein engineering studies have shown that when this group is removed the enzyme functions a million times less efficiently. Coincidentally, this is also the reduction in activity when comparing the specificity constants of hydrolysis and alcoholysis (transesterification).

By converting chemically Ser-221 to a cysteine we have been able to evaluate the role of this amino acid in organic solvents. Thiolsubtilisin (Ser-Cys 221) has been investigated for many years and it is known that the overall structure of the enzyme does not change upon synthesis. Thiolsubtilisin is, however, very susceptible to oxidation and must be treated with great care. Simply comparing the activity of subtilisin and thiolsubtilisin yields important information about how subtilisin functions in organic media. The calculation of the difference in the free energy of transition state stabilization between native and thiolsubtilisin. This free energy does not change significantly when the enzyme is placed in hexane, acetonitrile or dioxane, indicating that the active site of the enzyme must be protected from bulk solvent by either water or salt molecules in these environments.

Given that subtilisin utilizes the acyl-enzyme mechanism for transesterification, it is now important to elucidate the effect of mechanism, whereas in organic solvent alcoholysis reactions can be catalyzed. The mechanism is well understood in aqueous solutions, the structure of the enzyme has been determined, and subtilisin is readily available. Furthermore, it is possible to perform site-directed mutagenesis on subtilisin since the gene encoding the protein has been cloned previously. The mechanism of the enzyme in organic solvents has been investigated before.

During the acyl-enzyme mechanism a covalent bond is formed between the serine-221 hydroxyl group of the enzyme, and the substrate. Protein engineering studies have shown that when this group is removed the enzyme functions a million times less efficiently. Coincidentally, this is also the reduction in activity when comparing the specificity constants of hydrolysis and alcoholysis (transesterification). By converting chemically Ser-221 to a cysteine we have been able to evaluate the role of this amino acid in organic solvents.

Thiolsubtilisin (Ser-Cys 221) has been investigated for many years and it is known that the overall structure of the enzyme does not change upon synthesis. Thiolsubtilisin is, however, very susceptible to oxidation and must be treated with great care. Simply comparing the activity of subtilisin and thiolsubtilisin yields important information about how subtilisin functions in organic media. The calculation of the difference in the free energy of transition state stabilization between native and thiolsubtilisin This free energy does not change significantly when the enzyme is placed in hexane, acetonitrile or dioxane, indicating that the active site of the enzyme must be protected from bulk solvent by either water or salt molecules in these environments.

Given that subtilisin utilizes the acyl-enzyme mechanism for transesterification, it is now important to elucidate the effect of different solvents on the relative rates for acylation and deacylation of the enzyme. Clearly, if we are to optimize catalysis in extreme environments we must first understand which steps in the mechanism are rate limiting, and what the rate of each step is. This is not a simple procedure for the use of enzymes in hetrogenous biocatalytic reactions. There are number of different approaches which may be usedto determine the microscopic rate constants for alcoholysis. Importantly, each of these methods has been used before to either study enzyme-catalyzed reactions in water or classical hetrogenous catalytic reactions. As such the methods themselves are novel only in their application to this system.

The use of steady-state isotopic transient kinetic analysis (SSITKA) for determination of microscopic rate constants of the subtilisin-catalyzed alcoholysis of N-acetyl-L-phenylalanine ethyl ester by propanol. It is possible to perform steady state experiments with radiolabelled substrates in order to investigate each step of the reaction. The main drawback of this approach is that before it is possible to determine actual rate constants one must first quantitate the role of diffusion. For this approach to be successful it is necessary to be able to saturate the enzyme with both substrates, and the rate constants for acylation and deacylation should not differ by more than an order of magnitude.

Under these circumstances determination of kcat and Km for each substrate, in the presence of varying concentrations of a second added nucleophile allows the determination of the microscopic rate constants for acylation and deacylation as has been described previously for chymotrypsin in mixtures of water and butane-1, 2-diol.

Solvent engineering is one approach for increasing activity of enzymes in non-aqueous media. Much can be achieved, however, utilizing the lessons learned from protein engineering studies in aqueous media. Just as it is possible to measure the strength of interactions within a protein in aqueous media via site-directed mutagenesis, it is possible to evaluate the important criteria for rational control of activity by measuring the activity of engineered proteins in extreme environments.

Future Development of Mechanisitic Enzymology

An important future development is that much of enzymology can be simulated using methods based on essential physical laws found in common with computational chemistry, materials science, and combustion. However, there are many challenges to the simulations of these biological processes that arise from the large molecular sizes and long timescales relevant for biological function, or the subtle energetics and complex milieu of biochemical reactions. These challenges are greatly amplified by the need for high-throughput efforts to annotate the large number of genome sequences to include an understanding of their biological mechanisms and functions.

Some of these same computational methods and bottlenecks arise in the evaluation of the binding affinity of drugs to specific targets, but the task of the drug designer is further complicated by the need to identify a small group of compounds out of a virtually limitless universe of combinatorial possibilities.

In rational ligand-docking approaches, the interaction between two molecules is evaluated by computer simulation to quickly identify compounds that bind to a target with high affinity and specificity. Enormous chemical databases must be rapidly sifted through in order to identify a small number of candidate drugs which can be further evaluated by in vitro assays and animal studies.

Using this technique, it is feasible to evaluate hundreds of thousands of potential compounds within a matter of months or weeks, and model customized features such as improved binding geometry and pharmacokinetic properties. While rational, computer-based methods represent a quantum leap forward for identifying drug candidates, substantial increases in compute power are needed to allow for both greater selection sensitivity and genome-scale modeling of future drugs.

Reaction Mechanism of Malate Dehydrogenase

The minimum energy surface and reaction pathway for the interconversion of malate and oxaloacetate catalyzed by Malate Dehydrogenase (MDH) was simulated using semi-empirical

QM/MM methods. Analysis of the energy profile shows that solvent effects due to the protein matrix dramatically alter the intrinsic reactivity of the functional groups involved in the MDH reactions. The enzyme effectively changes the reaction from an exothermic reaction in the gas phase to a nearly isoenergetic one in the protein-solvent environment of MDH.

The minimum energy profile was determined by 675 separate energy minimizations, consisting of 1000 steps of Adopted Basis Newton-Raphson each, using parallel computers at NERSC and ANL. Each processor was assigned an MDH model with a different and independent set of distance parameters that define the reaction mechanism in the enzyme. The resultant energy profile showed that the MDH enzyme reaction is sequential, with the proton transfer preceding the hydride transfer.

In addition to giving a detailed mechanistic description of the MDH reaction, computer experiments with a QM/MM approach can also provide insights into the reasons that MDH is able to effectively catalyze the interconversion of malate and oxaloacetate. In particular, examination of the reaction profile shows that electrostatic effects in specific regions may enhance the transfer potential.

Damage-Recognition and Mechanisms for DNA-Repair Enzymes

Human apurinic endonuclease (APE) is a DNA-repair enzyme that plays an essential role in maintaining the fidelity of the DNA sequence by recognizing and repairing sites where a single DNA base has been lost from the sequence. The structure and function of APE has been the extensively studied, producing detailed kinetics data on a wide variety of substrates and an X-ray crystal structure for APE and related bacterial endonucleases.

Nevertheless, the mechanism by which APE recognizes the DNA is unknown. Numerous mechanisms have been proposed, including one that APE acts by inserting an amino acid into the abasis site or by recognizing the bare deoxyribose ring. Recent site-directed mutation results and kinetic studies using synthesized abasic DNA substrates show that these mechanisms for APE damage recognition are incorrect.

Instead a mechanistically simpler hypothesis has been proposed that involves the comformational flexibility of the damaged DNA after it is bound to the repair enzyme. Since there is no experimental method for directly monitoring such a mechanism, simulation is required to directly validate such an hypothesis.

Beyond the elucidation of the damage-recognition mechanism of APE, a detailed simulation will insight into the general mechanisms of the DNA repair enzymes. For example, a particularly interesting property of APE is that for certain damaged DNA substrates (those lacking the ribose ring), Mg^{++} ion is required for the catalytic activity, while for others it is not. MD simulations involving first principles force fields can help determine the location and role these ions have in the APE active site.

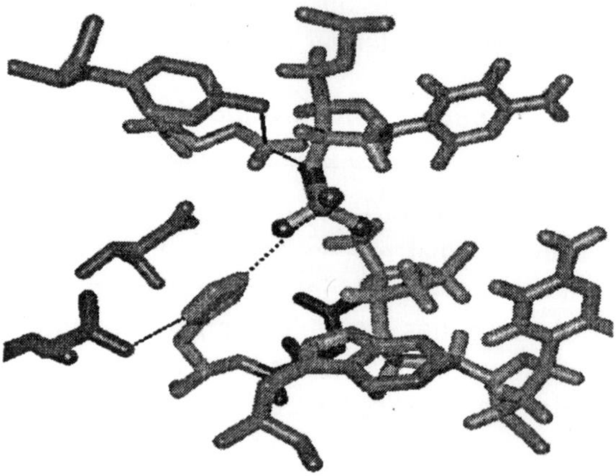

Figure 1. Damage-recognition mechanism of APE

Computational Enzymology

Modeling enzyme reactions at present involves two simulation approaches. One is semi-empirical or ab initio quantum mechanical (QM) methods that possess sufficient or even high accuracy for various biochemical properties of interest, but are currently limited to relatively small chemical systems and non-dynamic simulations. Another is classical molecular dynamics (MD), which simulates the motions of atoms in their chemical context for relatively large systems and long timescales, but with empirical force fields that often have insufficient accuracy, and altogether fail to treat the breaking and forming of bonds, that is especially important for enzymatic reactions.

The simplest level of ab initio QM simulation is the Hartree-Fock (HF) method. This method produces very accurate bond lengths and angles and reasonable reaction energies. Potentially more accurate structures and reaction energies can be determined with Density Functional Theory (DFT) that shares HF's favorable scaling properties. Promising new algorithms, such as the MP2 method, should allow for very accurate energetic calculations on the chemically significant segments of many biochemical reactions. However, for certain properties, such as reaction barriers that are particularly important in non-equilibrium biochemical processes, more sophisticated QM methods such as CCSD and CCSD(T) may be required.

If we consider the series of theoretical models, HF or DFT methods, MP2, CCSD, CCSD(T), for a given size basis set, and varying molecular size, M, then in the simplest analysis their computational requirements scale as M^4, M^5, M^6, and M^7 respectively. However, recent research has contributed to a rapid breaking down of the computational bottlenecks in HF, DFT and MP2 calculations.

Two steps are involved in one HF/DFT energy and derivative calculation. The first step is the construction of the effective one-electron Hamiltonian matrix, usually termed the

Fock matrix, given a density matrix. The second is the evaluation of a new density matrix, usually via the generation of new molecular orbitals or Kohn-Sham orbitals.

That HF and DFT methods naively scale as the fourth power of molecular size arises because of the evaluation of electron-electron interactions via four center two electron integrals. However, the number of non-negligible two electron integrals does not grow quartically with the size of the molecule, but grows as M2 when the molecular size is large enough (i.e. the two atomic orbitals (AO's) comprising each pair must overlap in order to make a distribution containing non-negligible charge). This realization, together with advances in the speed of two-electron integral evaluation (integrals are generated as they are needed rather than stored), combine to permit routine calculations on systems approaching the 100 heavy (i.e. non-hydrogen) atom range.

The next generation of quantum chemistry algorithms will exploit new theories and technology that will reduce the scaling requirements of the HF, DFT, and MP2 methods. For example, linear scaling in the assembly of the Fock matrix follows directly from the collectivization of distant electron-electron interactions via multipole expansions with controlled error bars known as Fast Multipole Methods. In the face of linear scaling methods for electron integral evaluation, the generation of a new density matrix via diagonalization that scales as M^3 will eventually become dominant for large molecular sizes. Current effort has been directed toward methods for updating the density and/or orbitals without explicit diagonalization, taking advantage of the fact that most molecules, such as proteins, the density matrix is spatially localized.

It is important to emphasize that DFT and existing functionals capture only certain types of electron correlation, and therefore the quality of DFT calculations are still under debate. The development of new DFT functionals is an active area of research. A potentially feasible alternative is the MP2 method that is the simplest wavefunction-based theory of electron correlation. In most current quantum chemistry programme packages MP2 scales as M^5; the M^5 scaling is a consequence of the formulation of MP2 using delocalized MO's which arise from standard HF calculations.

However, the MO's can be localized, and there has been some preliminary progress towards developing versions of MP2 theory based on localized orbitals. The "local-MP2" method scales only quadratically with molecular size, and comes to within a few percent of reproducing the exact MP2 energy with a given basis.

The simulation of certain enzyme catalyzed reaction mechanisms involving homolytic bond breaking (i.e. the breaking of electron pairs) or if transition metal atoms are involved in the active site, will require more accurate electron-correlated quantum chemical methods, e.g. coupled cluster (CCSD) that presently scale as M^6-M^7. The development of parallelized electron-correlated QM methods under the Scientific Simulation plan will allow the application of computational chemistry to these more complex enzyme mechanisms. For example a CCSD energy calculation should be feasible for a 40 atom system on a teraFLOP computer, which is sufficiently large to include a typical enzyme substrate and several catalytic amino acid residues.

Despite the great value of the static properties that can be calculated using QM methods, many biological processes are inherently dynamical. Such problems include processive reactions (DNA or protein synthesis), and processes such as macro- molecular conformational changes (DNA unwinding and allostertic enzyme regulation). Empirical force fields, without the inclusion of electrostatic polarization, cannot accurately describe the solvation of highly charged biomolecules and such force fields are inherently unable to treat bond making/ breaking reactions. Improvements will be made to these classical force fields, but a shift to quantum mechanical force fields (vide infra) will be required to achieve quantitatively accurate enzymatic simulations.

The primary advancement will be the merging of the QM and molecular dynamics methods to allow so-called First Principles MD, where quantum mechanical forces will be used to drive the classical motions of the atoms. Extension to dynamics simulations requires considerable methods development; the DFT force calculation must be converted to a linear-scaling method, and the entire molecular dynamics simulation must be implemented on a massively parallel computer.

Even with these improvements, first principles MD will not yet be feasible for long timescales and large molecular sizes. However, this capability will allow the solving of a large number of fundamental biophysical problems that have been inconclusively addressed by existing classical MD methods. These problems include the determination of the hydration structure of the DNA nucleoside bases; the energetic factors leading to DNA base pairing; the hydration of the DNA backbone and basic sites; and the role of polarization in the stability of protein a-helices.

Models and Algorithms for In Silico Drug Design

Docking methods are computational algorithms developed to both predict the three-dimensional structures of ligand-receptor complexes, and to evaluate the relative affinity or free energy of binding for these bound ligands or drugs. The need for improved docking and scoring methods is now especially acute given the future direction toward high-throughput annotation of genomes to generate new protein structural targets, combined with revolutionary advances in combinatorial synthesis of small molecule docking candidates.

The current combinatorial library paradigm is to design diverse drug libraries aimed at multiple but unknown targets or directed ligand libraries aimed at optimizing hits against individual targets. An inverted procedure is possible in which one or many libraries are screened on the computer against many targets to determine which libraries have the most desirable characteristics for which targets. This general approach can also include an "optimization" cycle where augmented libraries are scored against the best targets.

The fundamental attractiveness of this approach is that potential targets for all compounds can be addressed at a much earlier time and at much lower cost per compound, and is consistent with genome-scale drug design efforts. The basic challenge is how to improve the accuracy of the fundamental docking algorithms themselves while rapidly screening increasingly growing drug databases both in house and in the public domain.

Docking methods for geometric optimization of a candidate drug into a target active site is a solved problem when both the ligand and the target are treated as rigid objects. In some cases, limited flexibility is introduced by dividing the ligand into several rigid fragments that are docked separately. This represents the level of sophistication that is currently available from commercially available software packages. This approach is likely to identify, at most, one weakly binding compound per database of 100,000 chemicals, i.e. there are too many false negatives generated.

Introduction of full flexibility of at least the ligand for docking into a rigid target to refine the binding geometry has been shown to lead to better binding energetics, and therefore finding better drug leads in general. Flexible ligand and rigid target represents the upper limits of what can be attempted with current computational resources. When the peptide backbone and side chains of the target molecule are also treated as flexible, allowing the molecule to undergo locally induced conformational changes upon ligand binding, the resulting induced fit seems to be essential to understand ligand specificity.

Large-scale screening with full flexibility of both ligand and localized areas of the target is well-beyond reach with current computational resources, since only a few compounds can be screened in a realistic time. Essentially increased use of geometric refinement with at least full ligand flexibility, and ideally at least localized target flexibility, via standard optimization techniques for large libraries of drug compounds is an accessible and desirable goal in using future 100 teraflop computing.

Once a drug is geometrically docked, scoring of the binding affinity of a drug-receptor complex ranges from statistical multivariate equations that correlate X-ray crystal structural data of ligand-receptor complexes with experimental free energies of binding, to physically-based molecular mechanics approaches, to computationally intensive free energy perturbation methods.

Overall the rapidly calculated multivariate functions perform as well as the computationally intensive free energy perturbation calculations, with estimated relative binding free energies of about 2 kcal/mole, which corresponds to a binding affinity error ($\sim 10\Delta^{Gerror/1.4}$) of about 30-fold. The quality of these scoring functions provides a qualitative filter for ordering the binding affinity of drugs in large databases, but are not a reliable predictor of the most active drug molecules. Ranking drug affinity among many ligands for a given target is where multivariate functions work well, but it is unlikely that these existing scoring functions could determine specificity of a single drug against different receptors. It is even arguable that there is insufficient structural data for this case, so that multivariate statistical approaches are ultimately a dead-end.

Both molecular mechanics and free energy perturbation methods have the advantage of being based on physical interactions, so that alternative problems can be treated by the same approach. Molecular mechanics functions with solvent-accessible surface area descriptions for solvation have on average performed significantly worse than multivariate functions in the past, with binding free energy errors typically being 3 kcal/mol/. However, quite good correlation of the enzyme inhibitor activity of 33 inhibitors of HIV-1 proteases were determined by a purely molecular mechanics scoring function, and recent reported results for some new empirical force fields perform as well as correlated MP2 QM methods.

The cost of evaluating a 100,000 compound library against 100,000 gene products would take on the order of a full year on a dedicated teraflop computer. Certain large pharmaceutical companies have in-house databases approaching 500,000 drugs, and revolutionary combinatorial synthesis approaches are going to expand these databases even further. Add on top of that additional modeling accuracy requirements to better screen drugs, i..e. better empirical force fields, longer refinement stages in the calculation, and greater target flexibility, the search for better drug candidates will utilize well 100 teraflop capabilities on a sustained basis, and are inherently scalable beyond this projected computing goal for 2003.

High End Computing for In Silico Drug Design

The table below gives an estimate of the method and current computational requirements to complete a binding affinity calculation for a given drug library size. Going down the table for a given model complexity is a level of computational accuracy. The benefits of improved model accuracy must be offset against the cost of evaluating a model over the ever increasing size of the drug compound library brought about by combinatorial synthesis, and further exacerbated by the high throughput efforts of the genome project and structural annotation of new protein targets.

Modeling complexity	Method	Size of library	Required computing time
Molecular Mechanics	SPECITOPE	140,000	~1 hour
Rigid ligand /target	LUDI	30,000	1-4 hours
	CLIX	30,000	33 hours

Molecular Mechanics	Hammerhead	80,000	3-4 days
Partially flexible ligand	DOCK	17,000	3-4 days
Rigid target	DOCK	53,000	14 days
Molecular Mechanics	ICM	100,000	~1 year
Fully flexible			(extrapolated)
ligand			
Rigid target			
Molecular Mechanics	AMBER	1	~several days
Free energy perturbation	CHARMM		
QM Active site and	Gaussian,	1	>several weeks
MM protein	Q-Chem		

State-of-the-art quantum chemistry algorithms can expand the applicability of QM/MM methods to simulate a greatly expanded QM subsystem for enzymatic studies, or even estimation of drug binding affinities. An estimate of the cost of using QM methods for evaluating a single energy and force evaluation system of 10^4 heavy atoms, would require resources that can handle $\sim 10^{16}$ FLOPS; on a 100 teraflop machine this would require five minutes.

Parallel versions of these methods have been implemented and are available at a number of universities and DOE laboratories. On current generation teraflop platforms, QM calculations of unprecedented size are now possible, allowing HF optimizations on systems with over 1000 atoms and MP2 energies on hundreds of atoms. The presently available traditional-scaling ($\sim N^3$) first principles molecular dynamics code is running efficiently on serial platforms, including high-end workstations and vector supercomputers. The ultimate goal is to develop linear scaling quantum molecular dynamics code. It will be important to adapt these codes to the parallel architectures, requiring rewriting parts of existing programmes, and developing linear scaling algorithms.

Chapter 4

Biotransformation: Process Engineering Issues

Over the last five years a number of chemical companies have made significant investments to develop or acquire a strategic biotechnology capability. This reflects a growing recognition that biotechnology is one of key process technologies required by a world-class chemical company to maintain its competitive position. Many of the largest commercial processes are for ingredients for food products but increasingly other sectors of the chemical industry are realising the benefits that biological catalysis can provide for both large scale and smaller scale production.

This upsurge in interest and in new applications is most apparent in the fine chemical sector for the supply of chiral intermediates for the pharmaceutical and agrochemical industries and is being driven by dramatic developments in the technologies to identify, isolate and enhance new biocatalysts. There have also been a number of new applications of biotechnology for larger volume products. The existing and potential in-process applications of biotechnology in the chemicals industry include:

— biotransformations to introduce chirality

— biotransformations to change or introduce selective functionality

— fermentation for the cost effective production of a range of chemicals

Furthermore, biological processes offer the advantage of being operated under mild conditions of temperature and pressure, and with high selectivity leading to fewer by-products. This has provided opportunities to make compounds that are not readily made by conventional chemical synthetic routes or to replace existing manufacturing processes by easier routes with fewer steps and potentially lower environmental impact. Significant recent developments have also taken place in improving the productivity of bioprocesses. Biocatalytic reactions are now operated routinely at high yields in a variety of aqueous and non-aqueous systems to produce a wide range of chemicals.

Biotechnology is not the solution to all the problems of chemicals manufacturing. It is, however, one of the key enabling tools in the technology portfolio. Companies that do not have capabilities in biotransformations risk being excluded from some of the higher value segments of the market. For many companies the use of biotechnology in production processes represents a step associated with many risks and uncertainties. For those without any expertise in the biological sciences, these barriers may seem very formidable. Nevertheless, as the technology develops there is a range of opportunities to access and utilise it at relatively low risk. There is considerable university research underpinning the technology with a good record of technology transfer to industry. The UK is recognised as a leader in these areas.

Introduction to Enzymes

Biotechnology is not the solution to all the problems of chemicals manufacturing. However, it can be a key enabling technology in the technology portfolio. Biotechnology for the production of chemicals uses enzymes (a subset of proteins) to catalyse many classes of organic reactions and an increasing number are now used on an industrial scale. Several of the reactions catalysed are reversible and, by suitable adjustment of the process conditions, the enzyme may be used to catalyse a particular reaction in either direction. The following factors need to be considered for all enzymes:

— *Specificity*: Many enzymes show very high specificity in relation to substrate specificity (selectivity for a compound or group of compounds), regiospecificity (selectivity among similar groups (eg OH, O) on the same molecule); or stereospecificity (selectivity for one stereoisomer).

— *Cofactors*: To function, some enzymes need other materials present, ie cofactors. Two distinct types of cofactor exist. The first are simple metal ions whilst the second consist of complex organic molecules referred to as coenzymes. The more complex cofactors are expensive and this limits their application unless there is a practical scheme for their regeneration.

— *Biological Source*: Enzymes by definition are of biological origin. Specification of the source is often essential to define the particular enzyme. This is because enzymes with the same name from different sources often show differences in structure and properties.

— *Stability*: In general enzymes are more fragile than synthetic catalysts, although some can operate up to 60oC and function in non aqueous solvents. There is very little difference between the operating pH and temperature for different enzymes.

— *Control*: Enzyme catalysis is often susceptible to control by small molecules. For example, enzymes can be inhibited by the substrate, products of the reaction or by impurities present in the substrate. Understanding these control mechanisms can be important to process optimisation.

— *Mild Conditions*: In practice, there are several examples of enzymes enabling reactions to be run at temperatures lower than those with conventional catalysts and lead to better product quality, yield and reduced energy consumption.

For a company to develop a viable process using enzymes (biotransformation) they need to:

— *Identify the biocatalyst.* This involves screening and assessing commercially available enzymes and, possibly, the development of new biocatalysts. The development of new biocatalysts involves extensive screening of micro-organisms and other biological materials to identify the desired activity and then isolate the relevant enzyme or enzymes. Companies active in biocatalyst development maintain large collections of microbial strains and enzymes which are regarded as a key technology resource.

— *Source the biocatalyst.* In practice, companies can either purchase their biocatalyst or produce their own. Significant technical resources, involving various biotechnology skills, are required to develop and prepare biocatalysts in-house. The normal 'entry level' to access the technology is to use commercially available enzymes for single step biotransformations. However, it is possible to catalyse more than one chemical step using more than one enzyme in a cascade.

— *Process development.* A process for the desired biotransformation needs to be developed, scaled up and if necessary, integrated into the overall synthesis of the desired product. Optimising the processing downstream of the biotransformation is often just as important as optimising the biotransformation itself.

Health and Safety Good Practice

Like many substances used in industrial processes, enzyme preparations have certain health hazards, mainly as allergens. There is no scientific evidence that enzymes can cause problems by skin contact or ingestion, but inhalation as a dust or aerosol may stimulate the formation of antibodies.

The development of antibodies is called sensitisation and generally has no noticeable adverse effects. However, further exposure to the same allergen may result in symptoms such as itching of the nose and eyes, nasal congestion and sneezing through to coughing, tightness of the chest and shortness of breath.

There is also the possibility that some enzymes can cause eye, skin or mucous membrane irritation. Irritation is most likely to appear where there is perspiration (eg. hands, armpits, groin, feet and under a face mask).

The risk of developing an allergy to an enzyme or enzymes depends on two main factors:

— the susceptibility of the individual

— the duration and level of exposure.

Guidelines for Using Enzymes as Biocatalysts

These guidelines helping you to get the best out of enzyme catalysts when using powder and aqueous enzymes as biocatalysts. There will be exceptions and different conditions and circumstances for individual enzymes in specific situations.

Nature of Enzymes

Enzymes are naturally occurring proteins obtained from living materials. They are sourced from animal organs and tissues and plant materials. The majority of industrially used enzymes are from bacteria or fungi and can be used as catalysts (biocatalysts).

Enzymes can be structurally altered by various factors. If the deformation is minor, there may be a loss of all, or some, catalytic activity. This may be reversed by removing the deforming factor. More severe alterations usually result in irreversible loss of the catalytic activity.

Examples of reversible reductions in performance are:

- operating at a pH other than the optimum value for a particular enzyme (but not at extreme values)
- the presence of substances binding to the enzyme in competition for its substrate
- the absence, or inadequate amounts, of a substance required for the activation of the enzyme.

Examples of irreversible damage to enzymes are:

- extremes of pH
- extremes of temperature above the maximum tolerated by the enzyme
- the presence of irreversible inhibitory substances.

Information to identify the optimum conditions, the nature of any activators required and any specific inhibitors should be provided on the Product Data Sheet (PDS) provided by a reputable enzyme supplier. If the supplier knows the intended application, it is usual for them to supply additional documentation giving advice on operating conditions.

In addition, a reputable enzyme supplier will also supply a Material Safety Data Sheet giving information on safe handling, disposal of used material, action in case of accidental spillage and environmental issues. It is to be noted that some enzymes have the potential to cause allergic reactions and occupational asthma in susceptible subjects.

Operating Conditions

When adding an enzyme to a reaction system it is important to ensure that the operating conditions for the enzyme are stabilised first. This avoids damaging the enzyme with localised contact with added acid or alkali, heaters or steam injection.

Temperature

The most often stated benefit of using enzymes as catalysts is their ability to function in mild conditions.

Typically, enzymes as biocatalysts perform best at temperatures in the range 30 to 90°C. There are different maximum temperatures for different enzymes (obtained from the PDS) and some enzymes have been developed to function well at temperatures outside the general range.

The maximum recommended temperature will usually be 5°C below the value at which 'thermal death' begins. The rate of loss of catalytic activity is usually extremely rapid at temperatures above the stated maximum value. At temperatures below the maximum, the effect of changing the operating temperature will be to double the catalytic rate for a given quantity of enzyme for each 10°C rise, and to halve the rate for a similar reduction in temperature.

Generally, the best economic use of an enzyme as a catalyst will be at the highest practical temperature. If the reaction is not at the optimum pH, the damaging effect of heat is up to 50 times greater than at the optimum pH.

pH

The optimum pH given for an enzyme is usually determined under analytical conditions. Full activity may be expected 0.5 units below and above the optimum in the presence of large amounts of reactants that would normally be found in an industrial process. The enzyme's catalytic performance will decline in proportion to the deviation between the pH at which the reaction occurs and the optimum pH.

In an ideal situation it is recommended that the optimum pH for the reaction in question is found by experimentation and any effects on the enzyme activity determined. In many reactions, the products of the reaction alter the pH of the system. This should also be determined by experimentation so that steps can be taken to maintain the requisite pH within half a unit either side of the optimum throughout the reaction period.

The optimum reaction stated by the enzyme supplier is generally specific to a set of reaction conditions that can include the type of substrate used. For example, there are cases where the optimum reaction pH stated by the supplier is not the optimum when the enzyme is used for a similar class of reaction but with a different substrate. This is more likely to be the case when using the enzyme for an atypical reaction.

Quality of Purchased Enzymes

Activity decline

All enzyme preparations have a finite shelf-life and will decline in activity with time. This decline is minimised by the way the manufacturer prepares the enzyme and by the correct transporting and storing of the enzyme before use. A reputable manufacturer/supplier will ensure that the enzymes are transported according to their requirements and should, therefore, be fully within specification on receipt by the customer.

Assuming an enzyme is purchased for the functional activity stated and declared by the supplier, it is usual to receive material that conforms in every respect to the specification and the analysed activity stated on the certification. However, to ensure compliance with a stated shelf-life, most manufacturers will pack their product at a slightly higher activity value ('overfill activity') than that declared on the PDS but will not give the actual activity units.

The 'over-fill activity' may be up to 10% of the specified activity. Therefore, the 'overfill activity' may have some effect on the operational performance. Fresh materials may perform better than materials from older stock. This may also be observed by the user when a number of the same quantity batches are taken from ageing stock.

If the enzyme purchased is to be used for one or more of its ancillary (side) activities, not warranted by the supplier, it is unlikely that these will be found at a constant level in different batches of the enzyme. Also, the performance of side activity enzymes may decline at a different rate to the declared and certified enzyme. It may be possible to arrange with the manufacturer to re-standardise batches on the desired activity.

Powder enzymes

Powder enzymes, including granulated forms, generally have a longer shelf-life than aqueous enzymes. Typically, the supplier will warrant a shelf-life of six months at the purchaser's premises if stored in accordance with the supplier's instructions. The main requirements are to keep these products dry and away from direct heat (eg boilers, heaters, direct sunlight). Generally, powder enzymes are added to a system as a suspension and most suppliers recommend that they should not be added directly to the bulk system. To ensure the enzyme retains its designed activity, it is generally recommended that a prepared powder solution should not be kept for more than four hours before it is used.

Once the delivery packaging is opened, it is important to recognise that changes in ambient temperature will alter the humidity in the air and this can cause the moisture content of the powder or granulate to change. Stickiness, clumping and loss of activity will usually occur with changes in humidity. Careful resealing and avoiding changes in humidity in the storage area are important for maximising the shelf-life of the enzymes.

Aqueous enzymes

Generally, aqueous enzymes have a shorter shelf-life than powder enzymes (typically three months) at the user's premises. It is important that the storage instructions supplied by the manufacturer are carefully followed to maintain maximum activity. The enzymes must be stored away from direct heat sources and out of direct sunlight (which can cause UV damage to the enzymes if the containers are not completely opaque).

Mishandling of the container once it has been opened is the most frequent cause of reduced shelf-life or complete spoilage of aqueous enzymes. Water that evaporates from the liquid will condense on the upper surface of the container. Environmental contamination, introduced when the lid was opened, can cause microbial growth to occur in the condensate. If the condensate then falls back into the liquid, it will form a layer without stabilisers, and further microbial growth will occur that may affect the performance of the enzymes. Great care is needed if an aqueous enzyme is to be drawn repeatedly from a container. Similarly, care is needed to avoid these problems where continuous pumped dosing from a container is practised.

Ideally, aqueous enzymes should be delivered in full containers such that each container has either a whole batch quantity, or a number of complete containers that can be used for a batch operation.

Optimising Performance

Ideally, enzymes operating in a closed system will result in a reaction equilibrium between reactants and products. Getting the best out of an enzyme requires careful attention to shifting the thermodynamics towards product formation.

Many enzyme reactions can be directed to levels of above 80% product yield. The most effective pull towards products is to remove them throughout the reaction. This is not often practical. In some cases, a second reaction, often not enzymatic, is used to convert the product(s) to an insoluble form, separate them from the reaction for subsequent recovery treatment and return the reaction mixture to the reactor with additional substrate. Another example is to transfer the product to a second-phase liquid in which the reactants (substrates) are insoluble.

The sensitivity of the enzyme to substances in the reaction, other than the substrate, needs to be evaluated. It is important to identify the presence of inhibitors such as heavy metal salts and ions, protein denaturants, high ionic concentrations of normally benign substances and strongly acidic or basic conditions, and steps taken to avoid or minimise the impact of these substances.

The presence of unidentified or unsuspected substances can sometimes be indicated by making laboratory scale reactions under the intended operating conditions while increasing the amounts of enzyme incrementally. When carrying out such an analysis it is advisable to note the amount of conversion in a specific short timescale. The time is best set to ensure that there is no risk of the supply of substrate becoming limiting. The amount of conversion will be proportional to the amount of enzyme applied if there are no interfering substances. If the conversion is not linear to the enzyme amount, typically showing an asymptotic response, it is probable that there are problem substances present.

Enzyme Dosage and Cost

Enzyme dosage can be tailored to take account of some of the factors discussed above. For example, to obtain a short reaction time, for practical reasons, or to take advantage of higher enzyme performance at higher temperatures, it could be useful to increase the enzyme dosage. Furthermore, additional enzyme dosage may compensate to some degree when inhibitors are present that cannot be avoided or removed. However, the more enzyme used, the greater the enzyme cost.

It is often instructive to determine the 'half-life' of an enzyme. This is the reaction time, under the defined reaction conditions, when the reaction rate falls to 50% of the initial rate. It may be commercially useful to arrange to 'top-up' the system with more enzyme at this point and run on for another half-life or to consider this the practical end-point and adjust the operating conditions to reach this as quickly as possible.

Usually, the lowest enzyme cost system will operate at a low temperature (20°C down to 40°C), over a long time period

(20 - 60 hours), but this could allow microbial contamination and will probably require large reaction vessels.

Separation and Re-use of Enzyme Catalysts

Suitability of enzyme catalysts for separation and re-use: Separation and re-use of enzyme catalysts can be carried out only in processes where the enzyme is in an immobilised form (ie the enzyme is attached to a suitable supporting material or entrapped in the internal pores of a porous structured support). In both cases the aim is to have very large numbers of catalytic units exposed to the reactants and allow the enzyme to be removed from the product stream.

The principle purpose of having an enzyme in an immobilised form is to obtain a higher productivity or more economic process operation than can be achieved using soluble enzymes.

In some situations, the enzyme(s) of use to a particular process may be most effective if the micro-organisms producing them are themselves immobilised but not killed. This practice is often appropriate where the enzyme reaction requires the continued supply of a highly specific additional molecule (co-factor), coupled with regeneration and recycling of this co-factor. Using the whole microbial cell is usually the only economic way to maintain this type of reaction as the cost of obtaining sufficient amounts of the isolated co-factor is prohibitive.

Operating conditions of immobilised enzyme catalysts: The recommended operating conditions for an immobilised enzyme biocatalyst may be significantly different from those recommended for the same biocatalyst used in its soluble form. Immobilisation can confer substantial new characteristics to the catalytic function. In some cases, the biocatalyst will perform a reaction in the immobilised form that it will not carry out in its soluble form.

It is important to note that the half-life of a reaction has a significant impact on process economics when working with immobilised biocatalysts. The half-life is the reaction time under the defined reaction conditions when the reaction rate falls to 50% of the initial rate.

Furthermore, it is important to understand the test system that the biocatalyst supplier has applied to the immobilised biocatalyst to specify its activity. The operating conditions that will be used in an industrial plant are likely to be significantly different from those in the laboratory and could potentially affect the observed performance. The ability to perform the same test as the supplier is valuable in determining if the biocatalyst, as supplied, is the cause of an observed performance discrepancy.

Use of immobilised enzymes in batch reactions: In batch reactions, the immobilised biocatalyst is suspended in the reactants and kept stirred. The following factors need to be taken into account to achieve optimum performance:

— Reactants should be free from particulates as these may bind to the surface of the biocatalyst or block the pores of the biocatalyst.

— Diffusion gradients of substrates and products occur around the biocatalyst. If these are shallow due to inadequate mixing of the whole system, the overall reaction rate will be slower than predicted. To avoid this, the system should be stirred as fast as possible without causing abrasion from, or rupture of, the biocatalyst.

— As the reaction rate diminishes exponentially with time, a decision has to be made when to stop the process, recover the biocatalyst and use it to start a new batch. One method of doing this is to measure successive half-lives of the reaction. These half-lives will increase with time. When a half-life equals or exceeds the downtime, it is recommended that the process is stopped early and the biocatalyst recovered to a new batch. Subsequent cycles with the previously used biocatalyst will usually show longer and longer half-lives. For productive practice the maximum number of cycles of the same biocatalyst should be determined.

— The economy of a process could, potentially, be improved if a percentage of completely new biocatalyst is added to the original recycled biocatalyst. There will be a limit to the benefit of this practice as the bulk of the biocatalyst increases in the reactor.

— Washing the recovered biocatalyst before re-use is important. The steps required should be worked out with the assistance of the supplier and, if at all possible, include steps to remove any attached particulate materials and to reduce or eliminate microbial contamination.

At the chosen end-point of the reaction, the biocatalyst may be recovered by simple filtration, washed according to the supplier's guidance and then used in a further batch. To minimise the time for such a recovery process, it is generally important to select a biocatalyst support that gives a good distribution of the enzyme particles without them compacting together. This becomes more significant in elevated viscosity systems.

Biocatalysts can also be immobilised on magnetic supports. This approach allows the removal of the biocatalyst from the reaction by application of a magnetic field to the vessel (or parts of the vessel) or passing the whole reaction system through a magnetised collecting vessel.

Use of immobilised enzymes in continuous reactions: Immobilised enzyme catalysed continuous reactions are usually carried out with the biocatalyst being contained in a vertical column (with suitable retaining screening at each end) and the reactants running through the system. Downward flow has the potential to cause packing of the biocatalyst and reducing flow rates. Therefore, upward flow is generally preferred. The most cost-effective approach is to operate at the fastest flow rate that does not cause the biocatalyst to pack on the upper screen. Furthermore, columns are usually equipped with flow-through jackets to permit temperature control of the reactants within the column. As most biocatalytic reactions occur at temperatures below 100°C, warm water is the usual medium in the jacket.

The usual best practice with continuous column reactions is to operate several columns in series. The number and size of these columns will be determined largely by the production scale and, to a lesser extent, by the catalytic reaction rate.

To obtain as near as possible a steady-state production rate from a set of columns there are several factors to consider:

— Reactions are best carried out under the conditions recommended by the supplier of the biocatalyst. This will include the optimum temperature, pH and concentration of activating substances and the maximum flow rate. If the operating parameters are not known, it is recommended that they are determined by trials.

— Particulate materials in the reactants will reduce the flow rate and require an increase in operating pressure, which can become limiting. Particulates may also cause blocking of the catalytic sites.

— Products produced in column reactions usually need to be soluble to avoid blocking.

— Continuous columns may exhibit falling performance due to the growth of microbes within the column either in, or on, the biocatalyst matrix and support. Where the reactant stream is inherently likely to contain even small numbers of microbes, steps should be taken to remove these before the stream enters a column. It may be possible to sterilise the stream using heat, chemicals or filtration. This is especially important if the reactants include substances providing general nutrition for micro-organisms.

— It is important that the pH and temperature of the incoming reactants are fully adjusted before reaching the columns as there will be a relatively short contact time between the reactants and the biocatalyst in the columns. Any change of pH due to the desired reaction will reduce the performance of subsequent columns unless it is corrected in-line between each column.

— The half-lives of each column will be different. The effect of contaminants, inhibitors and microbial growth is usually the greatest in the first column.

— The use of more columns increases the optimisation and stabilisation of the overall performance.

Re-use of biocatalyst from spent columns may be possible, but methods for washing and activating the material need to be investigated. Unlike the use of biocatalysts in batch reactions, it is not generally beneficial to mix biocatalyst from a spent column with fresh biocatalyst.

Use of immobilised enzymes in membrane reactors: There are some applications of immobilised enzymes where the enzyme may be held within a reaction system by a semi-permeable membrane. This is sometimes applied where the substrate has a large molecular size and the enzyme, when in solution, is unable to pass through the membrane while the products of the reaction are small enough to pass through the membrane. A typical system of this type of application is the hydrolysis of polymers.

Attachment of an enzyme to the inner surface of fine tubular membranes packed together to create a reactor with a very large surface area is also possible.

Disposal of spent immobilised catalyst: As the cost of landfill disposal rises and environmental health issues continue to be integrated into industrial practices, the cost and practical issues of disposal become more significant. It may be helpful to consult the biocatalyst supplier regarding known acceptable methods of disposal of spent biocatalyst.

Engineering Issues for Biotransformations

The following four steps are required to design a successful biotransformation.

Characterisation

This leads to a definition of the reaction medium and the kinetic constraints and limitations imposed by the properties of the enzyme being used as a biocatalyst. Factors to be determined include:

— Substrate and product activity, water solubility, potential inhibitory and toxic effects, melting points and stability as functions of temperature and pH to ensure optimum enzyme activity conditions.

— The thermodynamic equilibrium between substrate and product to define any needs such as continuous product removal or alteration of temperature to attain the required level of conversion at equilibrium conditions.

— The rate of production of acid and base to ensure that the pH stays near the optimum for the enzyme.

— The production of water where water can limit the equilibrium conversion. A typical process where this is important is low-water esterification in which alcohols and carboxylic acids are combined to produce esters in the presence of lipases. The water needs to be removed.

— Gas consumption or evolution to ensure the type of reactor used is designed to allow adequate dispersion in the reactor without having too much gas-liquid interface that can lead to enzyme deactivation and foaming.

— Enzyme characteristics including availability and cost, the complexity of conversion and the number of steps employed, cofactor requirements and cofactor recycle costs.

— Kinetic measurements under conditions relevant to the envisaged process. This includes modelling to take account of the reverse reaction which will also be catalysed in reactions where there is thermodynamic equilibrium attainable between the reactant and product.

Mode of Operation

The mode of operation (such as batch or continuous) is based on the characterisation data, plant throughput and potential operating variables. Reactors most commonly used for biotransformations are the batch stirred tank reactor with, or without, substrate feeding and the plug-flow reactor. The latter normally involves a packed bed of supported enzymes or immobilised whole cells. The continuous stirred tank reactor is rarely used because they generally require higher levels of biocatalysts to achieve high conversions.

Operating Conditions

Reactant and product control. There is the possibility that the reactants and products can cause inhibitory or toxic effects on the enzyme or move unfavourably the reaction equilibrium to reduce conversion. These problems can be overcome by continuous feed of the reactants at the same rate as they are consumed and/or product removal as soon as it is formed.

Mixing. In reactions that produce acid or base, pH control is necessary. This usually requires the use of a stirred tank allowing rapid mechanical mixing in which the enzyme is used in a free or supported form. There is often a difficult design trade off in large reactors to balance rapid mixing with catalyst attrition losses.

Mass transfer effects. The inter-phase mass transfer needs to be controlled carefully where a second phase is present (gas, solid or organic liquid). The catalytic activity of some enzymes (notably lipases) is dependent upon the presence of an aqueous/organic liquid interface. Control of mass transfer between phases can be adjusted by:

— altering the agitation rate
— physically controlling viscosity (by using diluents or avoiding high biomass concentrations in whole cell biotransformations)
— adjusting the phase ratio in two-liquid systems (care must be taken to avoid stable emulsion formation or phase inversion)
— changing the particle size where solids are employed.

Handling Immobilised Catalysts

Immobilisation is desirable when whole cells have to be used, as it improves the ability to handle the biocatalyst using standard process equipment. Furthermore, immobilised whole cells are more stable, compared to free cells, when subjected to extreme conditions of pH and temperature and toxic reactants. However, the manufacture of such whole cell biocatalysts on a process scale is limited. The immobilisation of the biocatalyst on supports is often the most practicable option in operational terms and allows rapid and clean separation of the biocatalyst from the reactant mixture. Such options need to be carefully compared on cost and operational aspects with those that employ unsupported cells or enzymes. Immobilised biocatalysts are generally fragile and, when used in stirred tank reactors, careful design of the agitator is required to prevent excessive attrition.

Flowsheet Considerations

In general, the biotransformation operation will be part of a series of synthesis steps that mix biological and chemical operations. Consideration needs to be given for locating the biological steps within the overall flowsheet. In most processes, and especially continuous ones, the biotransformation should be located such that substantial adjustments to the physical condition of the process stream are not needed both upstream and downstream of the biological operation.

For example, these could include adjustment of pressure, temperature and/or pH or the removal and subsequent addition of toxic components. Where high temperature, pressure or reactant concentration reactions need to be performed, they should be carried out, as far as is practicable, before the biological step.

Accessing the technology

It is possible to identify three broad levels of use of biotechnology amongst companies in the Chemicals industry.

Level 1: On the first level, there are companies that are not using biotechnology in any of their operations. They may be unaware of the potential benefits, or be keen to either use or investigate the potential benefits of bioprocessing but do not know how to access the technology.

Level 2: The second level corresponds to companies that are employing biotechnology in their production processes using commercially available biocatalysts. These are normally enzyme preparations that carry out relatively simple chemical reactions, such as hydrolysis, and which do not need expensive cofactors. Importantly, the reactions can be conducted in existing reactors. These companies will not have the capability to produce their own biocatalysts.

Level 3: Companies on the third level are capable of preparing enzyme or whole-cell biocatalysts for their own specific chemical conversion requirements. Generally they will have a number of core technical capabilities, including screening, fermentation and recombinant gene techniques.

There are several barriers that have to be overcome if a company wishes to move into levels 2 or 3. A number of companies have successfully made the transition to level 2 and have demonstrated that it need not involve significant capital expenditure to introduce a biotransformation on a plant scale. In some instances dedicated plant has been designed and installed after the product and process had been proven.

In others biotransformations continue to be operated in multi-purpose plant.

Often the first barrier to be overcome is 'cultural', where the use of biocatalysts is regarded as difficult and quite different from the conventional chemical processes used in the company. This cultural barrier can also exist within companies already using biotransformations where the awareness and experience of the potential of biotechnology is confined to one part of the company. Tackling this cultural view usually requires education and training to demonstrate that biocatalysis can and should be considered together with conventional catalysis in process development decisions. It is important that each of the departments involved in introducing a process is considered i.e. management, R&D and production. As with all technical projects, success requires a project champion and the necessary management support and commitment.

A number of chemical companies have received considerable assistance from enzyme suppliers, which have practical experience in assisting other industries, to introduce enzymes in a manufacturing process. Initial concerns by production operators about the nature of biocatalysts, compared to 'conventional' catalysts, are quite widespread and the enzyme suppliers have been very effective in helping to address them.

The intensity of interest in the applications of biotransformations has resulted in the publication of a number of good and comprehensive reviews. These are a very cost-effective means of providing initial access and awareness raising for R&D personnel. Contact with Universities can often provide valuable access to available technologies and support for smaller and medium sized companies.

Identification of Specific Opportunities

The identification of specific opportunities to use biocatalysts and the assessment of their potential benefits needs an interdisciplinary approach. Most of the larger companies using biotransformations have initially introduced the technology to their operating companies through their corporate research and technology groups. For smaller companies, which may not have people with the necessary knowledge, an external technical resource can be useful. There are various sources of information and expertise on biotransformations and companies have successfully used a variety of approaches to access them. The use of external consultants, either University based or independent, is a common first step.

The United Kingdom has a very strong academic presence in the area of biotransformations. There are also consultants with industrial experience of applications of biotechnology. The major enzyme suppliers are willing to offer technical advice as part of their effort to market enzymes for use in this sector and are themselves an important source of information. Additionally, a number of UK Universities run short courses that are designed to train organic and process chemists in the basis principles of biotransformations and biochemical engineering. Some companies have funded or part-funded University projects in order to access and transfer technology.

The scale-up of a biotransformation from laboratory reaction to production operation will differ in detail from the scale-up of a classical chemical process. However the principles of process development and optimisation are familiar to most of the companies in the sector. As with classical chemical processes, each process is different and has to be optimised individually. There have, however, been recent moves to develop a coherent and generic approach to tackle process optimisation of enzymatic reactions and to help ensure that this is done as effectively as possible.

Where processing issues concern the physical handling of the biocatalyst, technical support is available from the enzyme suppliers. Many companies have the internal resources to solve the process scale-up problems for the implementation of their first biotransformation. The key process challenges, in some cases, do not relate to the biotransformation stage itself but to the subsequent product recovery and work up. Contract manufacturing companies can also be employed to custom develop specific processes.

The resources needed to develop a new biocatalyst effectively are substantially greater than those required to utilise commercially available enzymes. A significant level of expertise in a variety of biological disciplines is required to allow the screening of organisms for useful biotransformations and to develop and prepare the appropriate form of biocatalyst. The development and production of most biocatalysts, whether enzymes or whole-cells, now involves the application of modern molecular biology techniques. However the possibility of significant technical advance, which can be protected by patent, is extremely attractive and a number of companies have built up formidable strengths in this area.

As biocatalysis develops further and biotechnology techniques become more established, widespread and routine, it is highly likely that there will be opportunities for more medium sized and smaller companies to access this level. For some it may be appropriate to establish their own biotechnology capability. In most cases a collaborative approach is likely to involve lower costs and lower risks. Collaboration with an academic partner may be beneficial.

The role of the companies is to evaluate the biocatalysts developed, consider relevant process options and devise a suitable biocatalyst production protocol. Biocatalyst production can be carried out by a third party. By having some control over this stage, however, the company is in a much better position to obtain maximum added value from such a new process. Collaboration with other companies is also likely to be a practical approach to accessing this level of technology. In particular the smaller companies with a strong research base, including the new biotechnology companies, are potential partners.

A number of companies have been formed in the last few years that specialise in the discovery and development of novel biocatalysts for specific purposes. Nearly all such companies make provision to license technology and to enter collaborative projects as part of their overall business and financial plan. These collaborations are much lower risk than developing a biotechnology capability from scratch in-house. Inward licensing can be used to take larger steps and, given the nature of the market, is expected to be a practical option for many companies.

Stimulating the Right Reaction

Biocatalysis, in combination with conventional organic chemical synthesis and purification, can allow a significant reduction in the number of synthetic steps and increased production efficiency for synthesis of chiral compounds. These are compounds which occur as pairs, each composed of the same atoms but structurally arranged as mirror images of each other. Compound pairs are known as enantiomers, and although they have identical chemical compositions, they can have specific and quite different biological effects.

An example is the herbicide intermediate 2-chloropropionic acid, where there are significant economic advantages to using the chirally pure (S)-enantiomer (S-CPA). The traditional process for S-CPA involves the multi-step chemical conversion of (R)-lactic acid (produced by fermentation from glucose). In the Avecia biotransformation process, racemic (i.e. equal quantities of two enantiomers) CPA (a low-cost commodity chemical) is treated with a micro-organism containing an enzyme which specifically converts the non-active enantiomer (R)-chloropropionic acid (R-CPA) to (R)-lactic acid, but does not react with the desired S-CPA. The reaction mixture is acidified to precipitate the biomass, which is then filtered off. The S-CPA is extracted with solvent and purified by distillation.

To improve the performance of the biotransformation process, genetic manipulation of the micro-organism produced a strain with a tenfold increase in enzyme productivity compared with the original strain. Without this modification the economic production of S-CPA would not have been possible.

Scenarios in Past, Present and Future

Biocatalysis has developed to such an extent that enzyme-catalysed reactions have now become an established technique. Advances in technology and continued research have largely overcome the limitations associated with biocatalysis since its inception over 30 years ago.

Biocatalysis, in combination with conventional organic chemical synthesis and purification, allows companies to reduce significantly the number of synthetic steps and increase production efficiency particularly for chiral intermediates. Biocatalysis can be applied to a wide range of reactions. However, it is best suited to the synthesis of products that are difficult to produce via conventional organic chemistry compounds or where there is a requirement for specificity.

Drawbacks associated with biocatalysis in the past such as limited number of commercial enzymes, narrow range of useful transformations that enzymes mediate, enzyme instability and long development times can now be overcome.

A combination of new discoveries and techniques (eg directed evolution and mass automated screening) has helped to access novel enzymes. However, subsequent biotransformation processes at larger scale can be one of the most challenging stages from initial discovery to routine manufacturing. But the UK is fortunate to have a strong academic base that can address this important step and provide early lab-scale development. As biocatalysis develops further and biotechnological techniques become more established, widespread and routine, there will be greater opportunities for more medium sized and smaller companies to access this technology.

Novel Screening Methods

In essence, screening means searching for a particular gene coding for the target enzyme. The starting point for the process of search and discovery is the rich diversity of micro-organisms in nature. Usually, enzyme screening and selection strategies are based on knowledge of the application and the physical and chemical conditions under which the enzyme must operate. Therefore, selection for enzyme performance under the application conditions is an essential early step in the screening process. Access to a large gene pool is a prerequisite for a successful screening programme.

The last decade has witnessed spectacular progress in molecular screening, enabling the direct sequencing of nucleotides of a DNA strand. Already, there are thousands of gene sequences in public databases, although most relate to genes involved in the translation of the genetic code. Since the genetic code is universal, it is possible to predict the probable sequence of nucleotide based in the gene coding for the specific enzyme.

The most recent advance in designing enzymes is the realisation that, with advanced recombinant DNA technology, it is possible to accelerate evolution to a point where thousands of mutations can be created in successive generations. This is subject to either direct selection or to indirect selection by way of a rapid screening strategy. This approach has been termed 'directed evolution'.

These techniques generally produce numerous incrementally beneficial variants. A leapfrog technique that involves recombining the individual substitutions by shuffling their DNA has also been exploited. Classical mutagenesis is one of the most powerful techniques used to increase enzyme yields from micro-organisms. Genetic engineering allows the in vitro manipulation of DNA with the ability to reintroduce the modified DNA back into host cells. The genetic elements of genes, as well as their regulating elements, can be altered to achieve changes in enzyme expression. Cloning enzymes is another process carried out to get enzymes for a large-scale use. After having conducted a screening programme, cloning brings with it benefits such as significantly reduced cost of enzyme production and possible simplifications of processing as a result of using purer enzyme preparations.

Therefore, over the past few decades, we can say that advances in biotechnology have had an immense impact not only in the areas of therapeutics, pharmaceuticals and health care but also in industrial, food and feed applications. Recombinant DNA technology has proved to be an effective means for increasing the production of enzymes, including foreign or minor enzyme components, and for generating tailored enzyme compositions for numerous applications. Recent breakthroughs in screening methods and genomics have allowed greater and faster access to unique enzymes from culturable as well as unculturable organisms.

The ability to genetically engineer additional features into these naturally occurring enzymes and to accelerate the evolutionary process in a directed manner has provided routes to the design of unique enzymes that are tailored to a specific application. These technologies, coupled with the development of well-characterised generic hosts and optimised and controlled fermentation, recovery and formulation processes, have assured a reliable flow of safe, stable and cost effective products.

Chapter 5

Bioelectrocatalysis

Enzymes are able to catalyse direct (mediatorless) electron transfer. This phenomenon is also called bioelectrocatalysis. The major features that characterize mediatorless coupling, thereby differentiating it from the mediated type, are associated with: (i) electron transfer from the electrode to the substrate molecule, or vise versa, via the active site of the enzyme; and (ii) the catalytic nature of the whole process. In the case of mediatorless bioelectrocatalysis, the electron itself is a substrate for the reaction; enzymic transformation and the electrode reaction cannot be considered as separate reactions.

An important point characterizing bioelectrocatalysis is the catalytic removal of the reaction over-voltage. In contrast with mediated coupling, where the redox transformation of the mediator occurs without over-voltage (it is a reversible process), in the case of bioelectrocatalysis the enzyme catalytically reduces the over-voltage. In terms of energy profie, it reduces the energy of activation for the electrochemical transformation of the substrate into a product.

The ability to catalyse electrode reaction via mediatorless electron transfer has been demonstrated for a number of enzymes by different research groups worldwide. The application of the phenomenon of bioelectrocatalysis for the development of biosensors is associated with the ability of biocatalysts to form molecular transducers. The molecular transducer is a molecular layer formed by electrocatalytically active enzyme in intimate contact with the electrode surface, and this is capable of selective transformation of a chemical signal of the substrate into an electrical one. The electrical signal can be detected either by measurement of catalytically generated current flux (amperometric mode) or by measurement of the reduction of the reaction over-voltage (potentiometric mode).

Among electrocatalytically active enzymes, peroxidase and laccase are of special interest. These enzymes are able to catalyse important reactions for biosensor development, namely electroreduction of hydrogen peroxide [by peroxidase; eqn. (1)] and oxygen electroreduction [by laccase ; eqn. (2)]:

$$H_2O_2 + 2e^- + 2H^+ \rightarrow 2H_2O \qquad (1)$$

$$O_2 + 4e^- + 4H^+ \rightarrow 2H_2O \qquad (2)$$

Notably, the reduction of reaction over-voltage caused by these enzymes is as high as 400–500 mV. This factor results in a high potential for the use of these enzymes in the development of potentiometric biosensors. A peroxidase-based molecular transducer is based on the coupling of peroxidase attached to the electrode with an enzyme that specifically catalyses a reaction of analyte transformation accompanied by the formation of hydrogen peroxide.

Alaccase-based molecular transducer is capable of detecting oxygen concentration. This sensing principle is on the basis of the coupling of the transducer with a specific enzyme able to utilize or to form molecular oxygen as a result of the catalytic transformation of the analyte. The application of enzyme-catalysed direct electron transfer for the development of immunosensors has several advantages. For immunosensors, the detection principle is based not on detection of a substrate by a molecular transducer (as in the case of enzyme sensors), but on detection of formation}disintegration of the molecular transducer itself.

The enzyme with great potential for application for such sensors is laccase. Its substrates are atmospheric oxygen and an electron directly transferred from the electrode. This permits design of a virtually reagentless immunosensor. The basic principle of the detection of the immunointeraction is presented in Figure 1. The electrode is modified with antigen or hapten, and when laccase conjugate interacts with the antigen it forms a molecular transducer. When, owing to immunointeraction, the electrode surface is coated with a layer of laccase—antibody conjugate, a high electrode potential caused by the laccasecatalysed removal of over-voltage results.

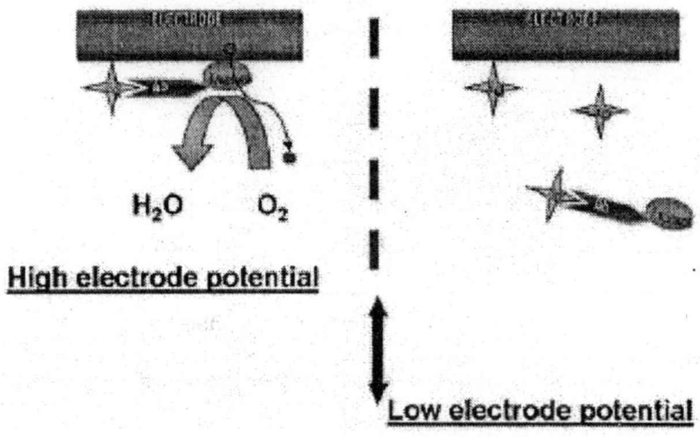

Figure 1: Laccase-based molecular transducer formed by immunointeraction

The electrode potential shift between the electrode before immunointeraction, and after the interaction, is approx. 400 mV. The potential shift itself is a characteristic of the area of the electrode surface covered by laccase immunoconjugate. Therefore the electrode potential shift can be used for quantification of the immunointeraction occurring on the surface of the electrode. It is notable that the conjugate in the bulk solution not attached to the electrode surface is unable to catalyse the electron transfer, and does not affect the electrode potential. Therefore phase separation is not needed.

A. Ghindillis, developed the study of application of biosensor development. Details are follows:

Electrode Preparation

Graphite Ink (EXP 741801) obtained from Ercon was deposited on a plastic strip (8 mm × 40 mm) employing a mask (0–1 mm in depth) to form a forking electrode of 4 mm-diam. with a connector 1 mm-wide, and then dried at 100°C for 4–6 h. The electrode body was then encapsulated with fast-dry enamel 720, obtained from Maybelline Inc., while the working electrode surface and a connector to the electrical circuit remained nonencapsulated.

The electrode was then pretreated by soaking in 3% (v}v) hydrogen peroxide solution for 20 min to create a hydrophilic surface. Immobilization of either the immunospecies or avidin was performed by physical absorption from the solution. The electrode was then incubated for 2 h at ambient temperature in a solution of trypsin inhibitor (0–1 mg}ml) in PBS to block free sites of non-specific binding. The electrode was then stored in PBS.

Measurement Set-up

The immunomonitoring instrumentation consisted of multiple test cells and sensing elements connected to a computer interface via a four-channel, high-impedance amplifier. A temperature controller was used to provide constant temperature during test measurements. A lap-top computer served for data acquisition and storage. The test cell employed was a plastic cuvette that permitted measurements in volumes as small as 300 µl. The thermostatically controlled cell holder had a lowfrequency vibrator, which served both to enhance analyte diffusion and to maintain a uniform temperature. This instrument permitted four different tests to be performed simultaneously.

Results

Different immunoassay formats have been developed for immunosensors on the basis of enzyme-catalysed direct electron transfer. An immunoassay based on the competitive scheme has been described for enzyme labels of both laccase and peroxidase. However, the sensing element based on the competitive assay requires a stage where conjugate is added before the addition of sample, thus this element cannot be designed as an all-solid immunoprobe.

In contrast, a sensing element on the basis of either a sandwich- or a displacement-assay format does not need components to start the measurement other than the analyte. This allows for the development of a reagentless immunoprobe.

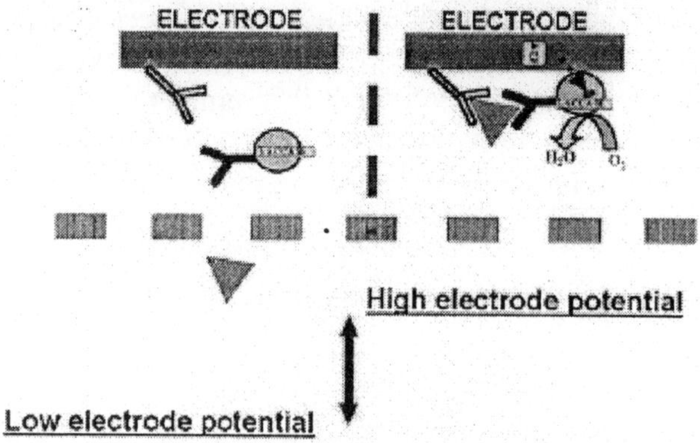

Figure 2: Schematic representation of sandwich-based immunosensing element

The sensing element based on a sandwichassay format requires a separation membrane impermeable to the immunoconjugate, but permeable to the antigen (analyte). The primary antibody is immobilized on the electrode surface, whereas the secondary antibody coupled with a laccase label is non-immobilized.

The diffusion of the analyte through the separating membrane towards the electrode results in formation of the complex antibody-antigen-antibody-laccase (Figure 2). This leads to an increase of the electrode potential, caused by the electrocatalytic removal of reaction over-voltage (eqn. 2). Such a sensor can be designed as a reagentless, separationfree assay probe.

The assay format on the basis of a displacement scheme has a similar advantage: the sensing element is an all-solid reagentless probe. In this case, the electrode surface is modified with immobilized, low-aænity antibody and preincubated with a conjugate of laccase–antigen to form a antibody—antigen–laccase complex on the surface of the electrode.

Free antigen (analyte) causes the displacement of the laccase-antigen conjugate from the electrode surface. This results in a decrease in the electrode potential that is proportional to the concentration of free antigen. The measurement can be performed either in the kinetic mode or by detection of the steady-state electrode potential, which corresponds to the immuno-equilibrium of the system.

Generally, the sensitivity of such an assay is less than that of the alternative ones. The lowest detection limit achieved by using insulin as a model analyte is 1 μM. To achieve the displacement of labelled antibodies by antigen-analyte, low-aænity antibody must be employed, limiting assay sensitivity. However, the simplicity both of the sensing element manufacture and of the assay procedure should be considered.

An alternative version of the displacementbased sensing element employs an immuno-interaction separated from the electrode surface. In this case, an antibody—laccase conjugate is modified additionally by biotin. This conjugate is bifunctional, having a site for immunointeraction with the analyte, as well as a site for specific interaction with an avidin-modified electrode. The electrode surface modified with avidin is able to capture specifically this conjugate due to the avidin—biotin interaction.

This permits reagentless quantification of the immunoconjugate on the electrode when immunointeraction occurs elsewhere (Figure 3). The rate of the avidin electrode potential increase is proportional to the concentration of the conjugate. This detection principle allows for the development of a reagentless detection with immunochromatography or immunofitration.

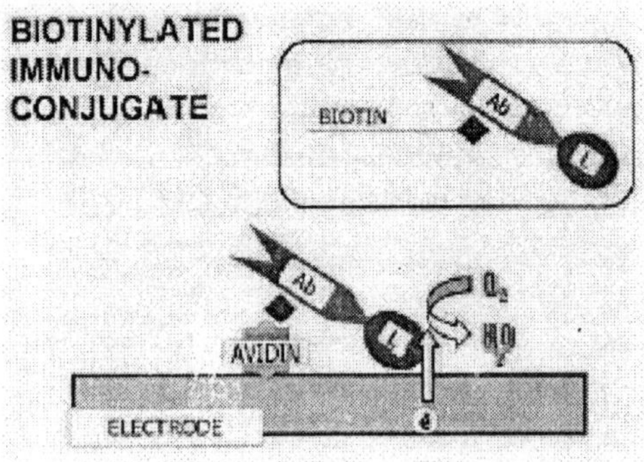

Figure 3: Signal transduction based on formation of biotinlabelled immunoconjugate on the surface of an avidinmodified electrode

For example, the flow of the conjugateand antigen-containing sample through an immunocolumn that has been modified with immobilized antigen is followed by the reagentless detection of the conjugate on the avidin-modified electrode. The resulting electrode response is proportional to the concentration of the analyte. In the case of immunofitration, where the analyte may be a large particle such as bacteria, the electrode signal is reversibly proportional to the concentration of the analyte.

The application of biotin-modified laccase ir. onjugates for the development of allsolid immunoprobes based on the displacement scheme has the advantage of high technological flexibility. The probe employs a separating membrane and an immunomembrane situated at some distance from the avidin-modified electrode (Figure 4).

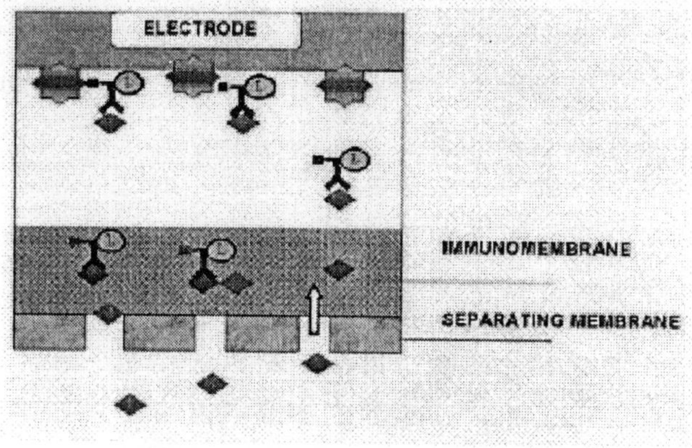

Figure 4: Schematic representation of a displacement-based sensing element employing a bifunctional laccase conjugate

Diffusion of the analyte through the separating membrane on to the immunomem-brane results in displacement of the laccase—antibody—biotin conjugate from the immunosorbent. The conjugate then diffuses towards the electrode and attaches to its surface by means of the biotin—avidin interaction. This results in an increase of the electrode potential, the rate of which is proportional to the concentration of the analyte. A low detection limit for insulin immunomonitoring below micromolar levels has been achieved.

Because the capacity and amount of the immunosorbent is variable, the technological flexibility of such an approach is high. It is also important that the same avidin-modified electrodes may be used with a variety of immunomembranes. This simplifies the opportunity for technological design of numerous immunoprobes for a wide range of analytes.

The advantages of the application of laccasecatalysed direct electron transfer for development of immunosensors are associated with: (i) the reagentless nature of the assay procedure; (ii) all solid format of the immunoprobes; and (iii) simplicity and low cost of fabrication, which allows for manufacture of disposable sensing elements. A potential for the practical application of immunoprobes on the basis of laccase-catalysed direct electron transfer is associated with:

— High Throughput Screening for new drug development;

— Individual quantitative immunoassay carried out by an untrained user (such as a patient);

— Application in combination with non-invasive or minimal invasive sampling systems for scheduled routine immuno-monitoring.

Chapter 6

Biochemical Reaction Engineering

There are many genes in the genome; 6,000 in a simple yeast and more than 30,000 for a human. In the past the genetic instructions could be crudely manipulated by techniques such as mutation achieved by, say, irradiation. Now scientists can alter the individual instructions of the genomic computer with molecular precision. Individual genes can also be snipped out and moved between organisms or chemically synthesized and inserted into cells. A relatively easy activity these days is to move one gene, say for a human protein, to a simpler organism.

If we are to maximise the benefit and minimise any risks this ability to replace the blunderbuss of crude methods by the precise rapier of molecular knowledge is crucial. An example shows what these techniques have already achieved. One of the great killers of the developed world involves a blockage of an artery. The region served by the artery's nutrient and oxygen supply will quickly be starved and irreversible damage will occur.

The human system produces an enzyme that selectively dissolves blood clots. However, the patient with a blocked artery does not have enough to cope and the levels in human plasma are too low to extract it from blood donations. The human enzyme, tissue plasminogen activator, can be produced using genetic engineering techniques. After very careful purification the enzyme specified by the gene is given by injection in the ambulance or casualty centre and saves lives —more than a million patients have been treated over the last 10 years.

These so-called biopharmaceuticals often represent the first real hope of dealing with previously untreatable conditions. Even when a protein could be obtained from human tissue, it can be much safer to use a genetically engineered material. Children who would be unable to develop to the normal height range can receive growth hormone from pure genetically engineered cells.

Haemophiliacs can now obtain the therapeutic protein, Factor Eight, they need to prevent bleeding without the risk of viral contamination from donated human plasma sources. In the era of AIDS and CJD this is a wonderful advance for these patients. Recently a new protein-based medicine has been produced which shows real gains in the treatment of arthritis.

Challenges to Biochemical Engineers

One of the earliest challenges was to address a potential threat to the large scale processing of proteins, which would have made the preparation of enough material for patients extremely difficult. These are very complex molecules as the simulated image of one of the smaller catalytic proteins, lysozyme, shows.

A molecule of the enzyme, lysozyme, in the form of a computer model. A linear gene codes for a linear sequence of protein building blocks, the amino acids, and the chain of fused acids is subsequently folded into this globular shape. It is mostly held together by non-covalent forces and one might expect it to be very easily damaged by the forces that occur in processes to purify the material.

The globular structure of proteins tends to unravel at airliquid interfaces and any intense mixing will bring proteins to such interfaces at a high frequency. Once this was realised engineers could be careful to design out the combination of intense mixing and interfaces. With such issues resolved the large-scale production of therapeutic proteins from genetically engineered cells became a realistic possibility.

Two particular biochemical engineering challenges are now facing us. The first is to cope with the even more complex and delicate materials that are emerging as potential medicines. The second is to increase the speed with which they can be brought from initial discovery through to the patient and to cut the cost of making them. Without radical new approaches the greater complexity will be associated with higher costs and already governments are very concerned with how to satisfy all the needs of patients.

Today a single injection of tissue plasminogen activator to dissolve a clot costs £1500. A year's treatment of a human therapeutic protein for multiple sclerosis costs £10,000 and because of this, treatment has effectively been rationed in the UK. Each such medicine takes an average of about 10 years to bring through from discovery to use, costing an average of about £400 million for each. The genetically engineered enzyme, tissue plasminogen activator, mentioned earlier was synthesised by a human gene instructing a simple cell to make it. Now scientists and engineers are moving to use the gene directly for therapy.

An electron microscope image of a gene contained in a loop of DNA called a plasmid. The great size of the plasmid causes the gene to be very sensitive to mechanical forces that occur in process equipment. If the pure gene could be placed efficiently and safely in a patient under precise control, it would not be necessary for multiple sclerosis or diabetic patients to inject themselves many times a year with a therapeutic protein.

The host of dreadful and incurable genetic diseases could be addressed and there would be an even better chance of dealing with the scourge of cancer by instructing the human system to return to normality. It now seems clear that such

DNA can also function as a vaccine and major efforts are being made to address diseases such as malaria and AIDS for which no vaccine yet exists.

Gene Therapy

The technique was to remove bone marrow cells and add the gene. This is a painful and laborious method and we need to have simpler methods of inserting the gene. There are many challenges to achieving this. One which faces engineers is that the gene which specifies a protein is a much larger entity than the protein it codes for and its less compact structure also make it more delicate. At present such problems are assessed by large-scale pilot plant studies. However, quite apart from the high costs of such work, the cumbersome and time consuming nature of large-scale trials poses a particular problem with new medicines.

Their development is a tortuous business because of the extended safety testing which is so essential. It means that, until clinical promise is quite well established, costly large-scale development is too risky. However, the dilemma is that once clinical promise is becoming clear the pressure to get the new medicine to patients is overwhelming in order that the company making it can obtain a return on the quarter of a billion pounds already spent. So, the time to develop an efficient process needs to be compressed. Unfortunately as complexity has increased, the delays at the pilot stage have tended to become greater.

Until now the complexity of biological systems has meant that these approaches have not been applied in this field. They involve the techniques of scale-down and of computer based modelling. It is widely acknowledged that refining conventional laboratory experiments often does not help to address the intrinsic constraints of full-scale engineering. This is acutely so for biological materials. Thus it makes sense to start with the likely large scale engineering limitations and to work backwards to small-scale mimics.

DNA as an example, centrifuging the disrupted cells of a genetically modified type to separate the human gene in its loop of DNA from cell debris will cause no problem in a lab centrifuge because the DNA is not subjected to significant forces. However, when the same material passes continuously through an industrial centrifuge, in order to process several hundred litres of material, the situation is very different.

There is instantaneous acceleration of every element of the liquid to the edge of the rotor with correspondingly large forces. The mechanical damage to DNA can be assessed using a small test cell with a fast rotating disc, which mimics the key features of the industrial rotor. Such analysis demands sophisticated techniques to account for the complicated stages through which the DNA breaks down. Using mathematical analysis it is possible then to link what is learned in this small device to what happens in the large industrial centrifuge.

It is close to the solid surfaces, that fluid is accelerated, where the most severe forces occur. Thus, by making the appropriate engineering connections it is possible to learn from a few tens of millilitres what will happen to hundreds of litres and this can be done quickly and at a low cost. Because many other biological complexes are also highly sensitive to mechanical forces, such methods are broadly applicable.

Successful scale-down from an industrial centrifuge to a small rotating disc device. The need for modelling the underlying engineering to make this connection applies also to understanding how all the possible ways of purifying materials such as DNA are to be assessed. The permutations of operations will be large because it can take a dozen stages to purify and formulate a biopharmaceutical to be ready for use. Until recently only a few of these stages had been modelled and for even fewer were these checked against large-scale experiments.

For example, when biological cells must be broken open to release products inside, the first instinct is to aim for complete breakage to obtain as much of the potential medicine as possible. However, the greater the rupturing force, the finer the cell debris and the more difficult it is to remove using industrial centrifuges, with serious consequences downstream. The process sequence can now be precisely modelled. Once "what if" questions can be asked at a computer with process models it is possible to avoid having to explore all the potential routes to a genetically engineered material at a pilot scale.

The first generation of genetic engineering was concerned with changing one gene of the thousands present. However, in principle there is no reason why much larger numbers should not be changed. This idea is being pursued in the new field of "metabolic engineering". In this way microorganisms can synthesise novel materials or make more of what is needed and less of what is not. The difficulty in manipulating many genes until now has been the sheer complexity of the outcome. Successful genetic modification of microorganisms demands a model of all the cell processes. This has parallels with modelling an industrial bioprocessing plant though the number of stages is much larger. Not surprisingly therefore biochemical engineers are becoming heavily involved.

Medicine for Good Health

If macromolecules pose processing problems, then human cells and tissue represent even bigger ones but they are increasingly being used to treat conditions such as knee injuries and to repair tissues following road accidents. These materials are living and therefore have to be supplied efficiently with nutrient, and especially oxygen. For cell and tissue repair, and equally for gene therapy, the material will often have to be tailored to the patient. It will need to meet the demands of the physician or surgeon and optimum use also will call for very close connection between scientist, engineer and industrial producer.

Biochemical engineers has been to understand how to produce the necessary fibres in quantity and with a precise control of their properties. We can achieve this by extrusion of fibre, rather like toothpaste, but in a very controlled way. We have also characterised just how far the cell aligning effects of the fibres extend. We are now looking at the repair of nerves and arteries using hollow fibres.

One of the challenges of "repairing" humans whose tissues are damaged is to create a support for building the replacement with human cells. Such supports need to be compatible with the human system and strong enough to be practically useful. Where relatively small numbers of patients have to be treated individually as they will in tissue repair and some gene therapy, there is a new consequence for biochemical engineers.

All the contact parts of conventional stainless steel systems used for making medicines have to be recleaned, sterilised and the outcome confirmed between each different material. That is simply too laborious and expensive for more patient-specific medicines produced in small quantities. Instead, plastic disposable process components will become essential as they have in medical laboratory procedures. The use of process disposable could also help start-up companies by substituting ongoing expenditure, as needed, for large expenditure at the outset. The fermenter vessel on the right is used for growing microbial cells from which biopharmaceutical macromolecules can be obtained.

Biocatalysis

Earlier engineering challenges of processing very large biological structures were described. But the biochemical engineer is also being affected by what is happening to the small chemicals used to make medicines such as antibiotics and herbicides that interact with living materials. Natural molecules can often be like right or left hands, which are not superimposable on one another. This derives from the fact that whenever a tetravalent carbon atom has four different groups attached to it there are two mirror image structures possible.

Until the 1990s most medicines were mixtures of both possible handed forms or isomers but after the horrors of thalidomide, where one isomer was safe and the other was not, it is recognised that pure isomer medicines are important. Making such isomers is more difficult and it is not necessarily just a single point in the molecule that must be right or left-handed.

A method using enzymes attached to beads for a key step in the production of modern penicillins as far back as the later 1960s. Today this method is employed almost universally. By using water immiscible organic solvents to dissolve chemical reactants, it was possible for a biocatalyst to act on water-insoluble chemicals from which new medicines are to be made. The inability of the two liquids to mix allows conversion without the biocatalyst being damaged by solvent. However, even though these developments have lead to some important industrial processes, many natural enzymes are not sufficiently stable to make them useful.

The new capacity to alter DNA can address this. A gene composed of DNA is like a necklace made up of thousands of individual beads, the nucleotides. There are now methods of shuffling the order of some of these nucleotides through the many possible alternatives. Provided a way can be devised of detecting a better enzymic catalyst coded by the new artificial genes, it is possible quickly to bring about molecular evolution that would normally take millions of years. The result is new biocatalysts that can allow new pure isomer medicines to be made.

Breadth of Vision

We believe that the education of biochemical engineers must stress breadth of vision, innovative thinking and flexibility of approach. Not only are these important in career terms in a fast changing world, they are vital attributes also for those who will bring to a practical outcome science with such far-reaching implications. Because the results will affect peoples' lives profoundly it is vital that the social and ethical issues are given sensitive attention.

From what has been described earlier it is clear that the quality of life of many people with medical conditions can be dramatically enhanced. A wide range of life threatening diseases that affect children, such as cystic fibrosis, as well as diseases suffered by older people will be addressable using the new approaches. The reason it costs about £350 million to bring each potential new medicine to the point of wide use is that the tests of safety are so stringent. The regulatory authority has to balance the benefit and the risk.

Biochemical engineers are continually concerned with ensuring that the methods, which were used to prepare the medicine, are the safest possible - a few virus molecules from a contaminated source are enough to make a valuable medicine a dangerous vehicle for infection, especially if it is given by injection. Biochemical engineers also are concerned with many non-medical processes.

Foods are complex mixtures of biological materials, human wastes are processed in what are really giant fermenters in which living organisms grow and detoxify the wastes. Many industrial wastes are also treated in this way and previously contaminated land is increasingly made safe by bioremediation in which special microorganisms are grown on site to break down harmful chemical residues. There is also progress towards using biomass (plants and particularly oilseed crops) as a source of fuel for vehicles. This will become increasingly crucial as non-renewable fuels become depleted.

When a therapeutic protein is produced by growing simple cells, such as a genetically modified yeast in a fermenter containing suitable nutrients, the organisms can be modified such that it is only able to reproduce in the special environment of the vessel. This "biological containment" is not possible if a microorganism or a whole agricultural plant is to be released to work in the open. The altered form must by definition be able to thrive in this natural environment. That means the general effect on the environment must be very carefully evaluated over an extended period.

As with medicine, in the final analysis there must always be a balancing of benefit and risk. It may be important to deal with a poisonous waste in such a way that a modified organism is a logical choice. Now, however, there are many more interested parties. The public just beyond the range of the toxic pollutant will need to be assured by good independent evidence that a genetically modified organism will cause no harm.

With biochemical engineers involved in major endeavours a concern with their environmental effects has to be a major priority. One feature of deliberate release is a particular issue with food produced from genetically "engineered" plants. Whereas the benefit of a powerful new medicine can be focused just on the intended patient, food from a modified plant will be eaten by very large numbers of people. It is against this background that so called genetically modified or GM foods have caused wide concern particularly in Western Europe. A better crop but one that enhanced say the irritating effects of some plant constituents in 1% of the population would be unacceptable.

Equally though, the precariousness of world food supplies is always a problem. If the efforts, say, to deal with global warming take longer than expected, major areas of high agricultural production may be lost. Equally problems of ground salinity with high water withdrawal for larger populations may make it necessary to seek salt-tolerant crops more effectively. In the Pacific Rim, where vitamin deficiency is a serious issue, a plant that could synthesise important vitamins would be invaluable.

Twelve percent of the world's cotton is now genetically improved by modern techniques. Future varieties could improve fibre quality to increase flame resistance and introduce popular colours into the plant so reducing the need for chemical dyes.

Applied Biocatalysis and Reaction Engineering

Enantiopure compounds production is increasing importance to the chemical and biotechnological industries. Bioorganic transformations are predestined to meet this demand due to their inherent regio- and stereoselective nature. Indeed, a growing amount of enantiopure chemicals for pharmaceutical purposes are being produced biocatalytically today, in contrast to the production of racemic bulk commodities in the past.

In this sense, biosynthesis needs to be understood as "chemistry by nature". The biological principles optimized over thousands of years experience a new renaissance when applied to technical asymmetric catalysis. Nevertheless, one key tool or prerequisite for their application is the appropriate technology, namely reaction engineering. The objective of the research group is to understand selected biological principles as well as processes in chemical terms, to transfer, apply them to and combine them with the traditional concepts of chemical, i.e., homogeneous catalysis. Simultaneously, the aim is to design and to develop novel reactors, i.e., engineered systems based on these biological principles.

Currently, in technical asymmetric synthesis different biological principles are already being applied without the scientific community at large being aware of it. In those examples namely, where additionally reaction engineering has been integrated, very efficient production processes were designed. This gives rise to high expectations for the future of a sustainable, technical asymmetric catalysis.

In nature complex molecules are selectively synthesized in networks and reaction sequences of biocatalysts, starting up from simple compounds. This principle is transferred to in vitro synthesis by setting up reaction sequences of different biocatalysts or even combinations of chemocatalytic steps with biotransformations. Here also new types of reactors like microreactors are evaluated for the application in the field of biocatalysis.

The second biological principle describes the properties of the biocatalysts itself. These are homogeneously soluble macromolecules, which are retainable by membranes. These properties are transferred to chemical organometallic catalysts in designing chemzymes. Based on kinetic and thermodynamic characterization of the reaction systems the limiting step is identified and appropriate reactors are developed. The overall aim is the design of new processes optimized in view of selectivity, catalyst con-sump-tion and space-time yield.

In Vivo Reaction Sequences

Biocatalytical hydrogenations are catalyzed stereoselectively with whole cells as well as with enzymes. For the diastereoselective production of 1,3- and 1,4-diols a process is developed utilizing whole cells from Lactobacillus kefir.

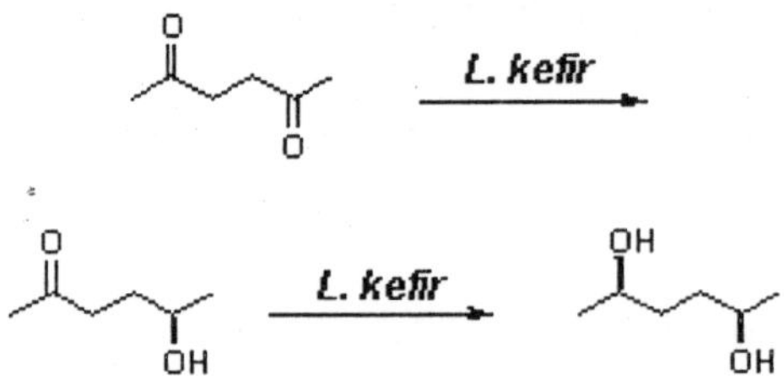

(2R,5R)-Hexandiol is synthesized in a two-stage reaction cascade with an *ee* and de >99%. As a result of reaction engineering the productivity of the biocatalyst could be increased by a factor 30 to 15 g/gcell dry weight. The synthesis is carried out in a continuously operated stirred tank reactor with cell retention. Space-time yields of 64 g L-1 d-1 are reached.

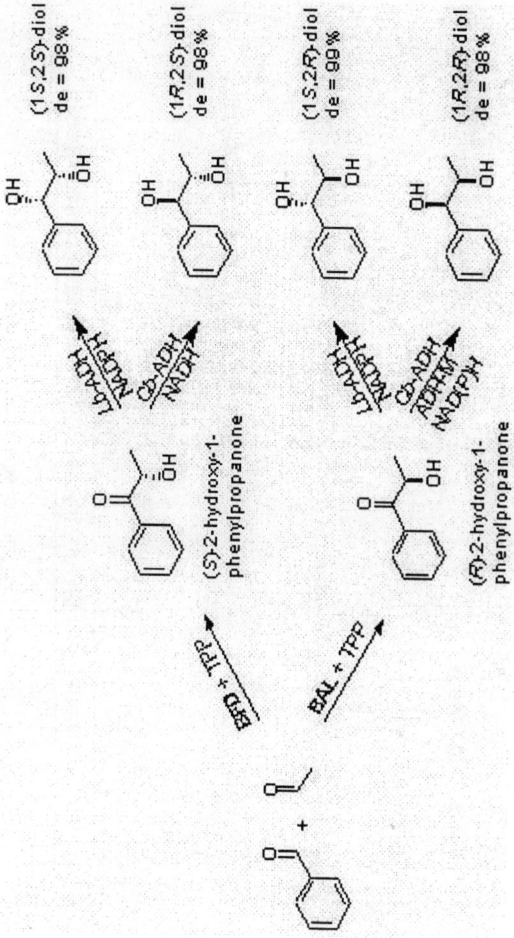

Figure 1: Implementation of reaction cascades

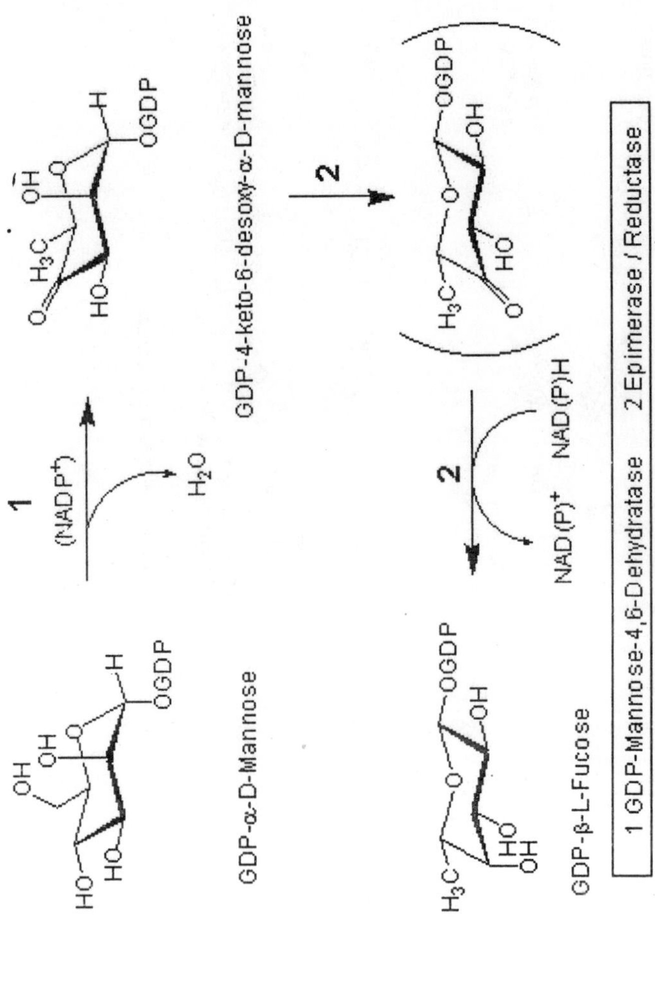

Figure 2: Enzymatic synthesis of GDP-fucose

Also isolated alcohol dehydrogenases are applied for the enantioselective reduction of multi-functionalized ketones. Most of the biocatalytical hydrogenations are carried out in multi-catalyst systems. The cofactor providing the reduction equivalences are either regenerated in whole cells by a sequences of enzymes or in case of an isolated alcohol dehydrogenase by the use of the second enzyme, e.g. formate dehydrogenase.

In Vitro Sequences of Biocatalysts

All projects dealing with the in vitro reaction sequences have the major aim of increasing the selectivity in common. On the basis of kinetic measurements simulations of different types of reactors are carried out to choose the best setup. Starting from simple aldehydes diols can be synthesized diastereoselectively by combinations of decarboxylases (BFD = benzoyl formate decarboxylase, BAL = benzaldehyde lyase) with alcohol dehydrogenases (ADH). By the combination of enzymes with complementary selectivities all four different diastereomeres can be synthesized.

Further examples for the attractive implementation of reaction cascades can be found in the area of oligosaccharide synthesis. By combination of a b-galactosidase with an a2,3-sialyl transferase the Fmoc-protected sialylated core1 structure is accessible. The synthesis of a core2 structure in a similar reaction cascade was optimized by means of a genetic algorithm. The selectivity for the product could be increased by factor 6.

Another example in this context is the enzymatic synthesis of GDP-fucose. By the action of two enzymes, a hydratase and a bifunctional enzyme (epimerase/reductase) in three sequential reactions GDP-fucose is synthesized starting from GDP-mannose.

Electroenzymatic Reaction Sequences

Often chemistry and biotechnology are complementary,

sometimes they are even competitive as in the production of fine chemicals. In view of atom- and energy-efficiency the combination of both fields is becoming more and more important. Electroenzymatic synthesis utilizes electrons as cheap reagents. The advantages of enzymes as enantioselective catalysts are combined here in reaction cascades with the waste-free electrochemical in situ generation or regeneration of reagents. By means of electrochemical catalysis reactants are either synthesized in situ or regenerated.

One example is the enantioselective oxidation of thioanisol. Here the oxidant hydrogen peroxide is produced in situ by cathodic reduction of oxygen. The enantioselective oxidation is carried out by a peroxidase, e.g. chloroperoxidase (CPO). The stability of peroxidase could be significantly increased in this electroenzymatic system. Here the overall aim is to develop upscalable electroenzymatic reactors.

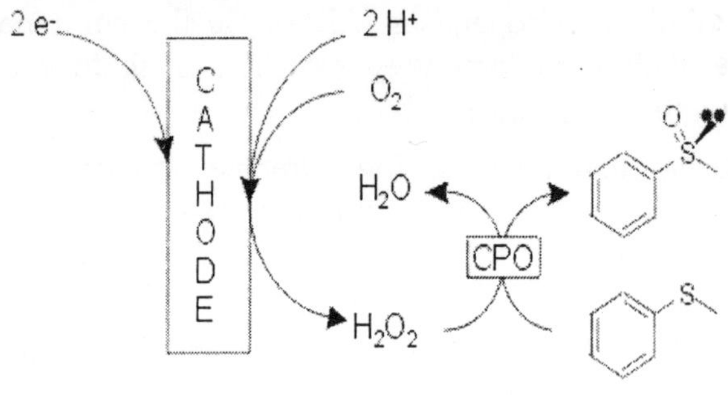

Figure 3: Enantioselective oxidation of thioanisol

Development of Chemzymes

Here principles of biocatalysis are transferred to enantioselective hydrogenations that are carried out with homogeneously soluble macromolecular organometallic catalysts (chemzymes).

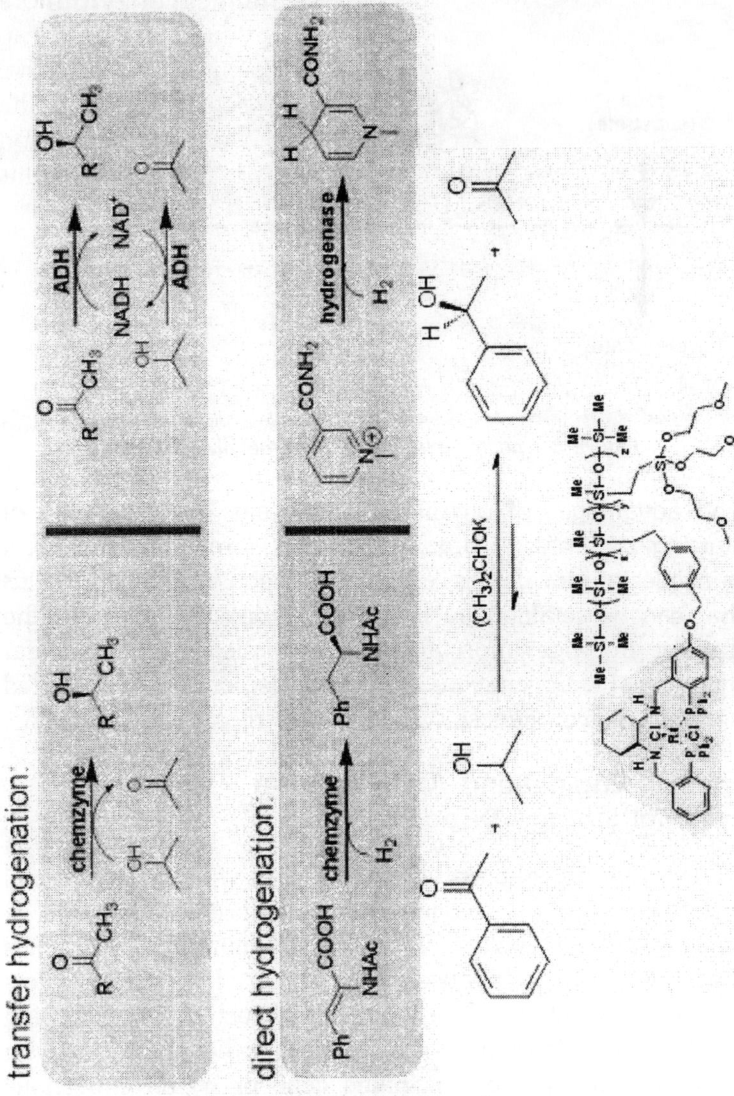

Figure 4: Hydrogenations

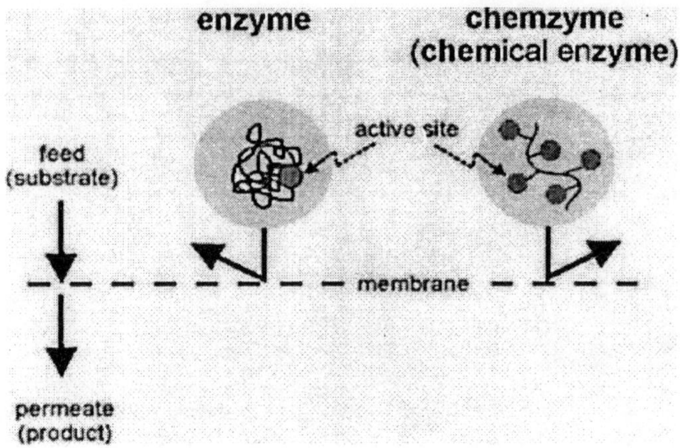

Figure 5: macromolecular organometallic catalysts

The advantages of organic catalysis, e.g. high solubility of starting material in organic solvents, are combined with principles of biocatalysis, e.g. homogeneously soluble proteins that can be retained by membranes. Special topics are the transfer hydrogenation and hydrogenation with molecular hydrogen. Hydrogenations are also carried out with an isolated enzyme, a hydrogenase, which utilizes molecular hydrogen.

Enzyme Engineering Application in Lysomal Storage Disease

In 1963, arylsulphatase A was one of the first lysosomal enzymes which was recognised to be deficient in disorders of sphingolipid metabolism. Since then a number of enzymes are now known to be absent in a variety of storage disorders as shown in Table 1. The treatment of such disorders arising from enzyme deficiencies can be approached in two ways : by correcting the malfunctioning gene or by an exogenous replacement of the missing enzyme. Both the treatments involve the delivery of appropriate molecules (DNA fragments or intact enzymes) to appropriate intracellular sites.

Table 1: The deffects in human lysosomal storage diseases.

Disease	Deficient hydrolase(s)
Tay-Sachs disease, A variant.	β - Hexosaminidase A
Tay-Sachs disease, AB variant.	β - Hexosaminidase activity with GM2 ganglioside as substrate
Pompe's disease	∝ - Glucosidase
Metachromatic leukodystrophy	Arylsulfatase A
GM1 gangliiosidosis	β - Galactosidase
Hurler's disease	∝ - Iduronidase

In enzyme therapy various approaches have been used to compensate for the defective enzyme. One of them has been orqan transplant, as in the case of Fabry's disease where it has been possible to transplant a normal kidney in a patient. This approach has shown considerable success in alleviating the disorder. Another approach is the transfusion of normal plasma which has been used in the treatment of Fabry's disease and Hurler's syndrome. Though this method had some initial success, its curative effects were short-lived due to a rapid inactivation or removal of enzyme from circulation. Before reaching the site of action these foreign substances provoke adverse immunological reactions, undergo proteolytic degradation by blood enzymes, and are even taken up by RES and nondiseased tissue preventing thereby their delivery to diseased tissue in sufficient amounts.

As a Drug Carrier

In the search of an appropriate carrier for such therapeutically active substances it was felt that various problems of enzyme therapy might be circumvented by their entrapment in liposomes. Liposomes, which were first developed by Bangham, are biodegradable lipid vesicles with concentric lipid bilayers alternating with aqueous compartments. Water soluble substances like enzyme can be entrapped in these aqueous compartment. In vivo experiments have shown that the activity of an enzyme in liposomes remains latent while they are in circulation and much of this enzyme activity is eventually recovered from the hepatic lysosomes, indicating that the liposomally encapsulated enzyme is protected from the environment.

Another advantage in the use of liposomes is that they can be targetted to the diseased tissue. Kinsky suggested that liposomes can be prepared by incorporating glycolipids such as cerebrosides into the matrix with the carbohydrate moiety of the glycolipids exposed on the surface of the liposomes. This led us to suggest that modification of the liposomes by incorporating various glycolipids having nonreducing terminal sugars would facilitate the targetting of the liposomes to different tissues carrying different terminal sugar specific receptors.

Ghosh et al showed that liposomes containing asialo GM1 ganglioside, when administered intravenously accumulated in the liver. The observed uptake was dependent on the density of the galactose residues on the liposomal surface, and was inhibited by competition with asialofetuin. They further observed that β - gal liposomes accumulated preferentially in the hepatocytes as compared to ∝ - gal liposomes indicating that the hepatic receptors are specific for β - gal containing liposomes.

It has been shown that when such liposomes with the entrapped enzyme are administered intravenously into mice. They are taken up predominently by the livers of the animals. Subcellular fractionation shows a considerable proportion of this to be in the lysosomes of the cells. This is of significance as quite a large number of inherited enzyme disorders arise from the deficiency of lysosomal acid hydrolases. All these factors suggested that liposomes would be ideal carriers for therapeutic enzymes.

Gregoriadis and Buckland tested the efficacy of liposomes as enzyme carriers in a model for storage diseases. When Chinese hamster fibroblasts were exposed to sucrose, there was an accumulation of sucrose within their lysosomes due to the absence of invertase in these cells. Subsequent exposure of these cells to liposomally encapsulatd invertase led to the degradation of the stored sucrose indicatinq that invertase was incorporated into the cells.

In another study, Colley and Ryman developed a model storage disorder in rat liver by accumulating dextran. In the absence of the enzyme 'dextranase', dextran is not degraded in mice. However, administration of liposomally encapsulated dextrance led to a significant degredation of stored dextran. Liposomes, therefore, appear to be promising as vehicles for the administration of enzymes. Gregoriadis and Ryman observed that although liposomally entrapped amyloglucosidase could be selectively delivered to the liver lysosomes, but once there it is very rapidly catabolised. This is true for other enzymes proteins as well.

The rapid degradation of the administered enzyme severely limits its therapeutic potential and thereby the efficacy of enzyme therapy in controlling such disorders. Work therefore started in several laboratories to find ways of enhancing the stability of the administered enzyme in the lysosomes. In order to do so successfully it is first important to understand the nature of lysosomal hydrolases, which despite being proteinacious in nature are remarkably stable in the acidic milieu of the lysosomes. These enzymes have some properties in common : (a) They have in general an acid pH optimum, (b) they are generally resistant to autolysis and (e) they are glycoproteins. The glycoprotein nature of the lysosomal hydrolases and their resistance to autolysis lead us to investigate the importance of the oligosaccharide part of these glycoprotein enzymes.

However, removal of the terminal sugar lead to a markedly reduced survival time in the circulaton indicating that it had a role to play in the in vivo circulatory life of the enzyme. The glycoproteinic nature of the lysosomal enzymes as well as their resistance to autolysis led us to suggest that it might be possible to prolong the *in vivo* survival of enzymes by their covalent modification with polysaccharide.

Covalent Modification of Enzymes with PEG and DEXTRAN

With a view to enhance the stability of enzymes *in vivo*, various investigators covalently modified a number of enzymes with hydrophilic polymers such as polyethylene glycol (PEG) and dextran. They have shown that PEG and dextran reduce the immunogenicity of a number of enzymes with a minimally deleterious effect on their catalytic functions. This is particularly important where repeated treatment with the enzyme is required over a period of time. Streptokinase, which promotes autolysis by activation of the fibranolytic system of the human blood and is clinically used in the treatment of thrombic states in man, is one such enzyme.

Being an enzyme of bacterial origin, it is strongly antigenic. Koide et al reported a complete loss of antigenicity with little or no effect on the activity of this enzyme after its modification with PEG. Such modified enzymes also exhibit enhanced circulatory life as compared to the native enzyme. Various factors have been evoked to account for this increase in circulatory life. These include loss of antigenicity, increase in size of small peptides and hence a decreased clearance by glomerular filteration in the kidney.

The hydrophilic coating also prevents the binding to and removal of the enzyme by "opsonins" , which are known to have particular affinity towards the hydrophobic reqions of the molecule. In vitro studies have shown that the modified proteins exhibit enhanced resistance to proteolytic degradation. This could well be responsible for the increased circulatory life. Such modified enzymes are increasingly finding clinical use in the treatment of various deficiency disorders.

Chapter 7

Metabolic Channelling in Enzyme Systems

In the beginning of modern biochemistry, enzymologists have studied enzymes in vitro. This approach to understanding cellular behaviour is based on the belief that the phenomena observed in cells can be attributed solely to the properties of the cell components. The quest then has been that of isolating the thousands of different enzymes in the living world and studying their physico-chemical properties in vitro, with the implicit assumption that after this knowledge had been attained one could somehow "reconstitute" the properties of the cell, if not in practice at least in principle.

Channelling

The structure of the cytoplasmic matrix, there is a considerable amount of evidence for enzyme-enzyme interactions, both in this compartment and inside other cellular organelles such as the mitochondrion, and within prokaryotic cells. These interactions could of course simply be a consequence of proximity effects if enzymes are part of the MTL or are adsorbed to it in the cytoplasmic compartment, or because the enzyme concentration is high (as in mitochondria and prokaryotes). However, a very interesting and important set of enzyme-enzyme interactions, for the existence of which there is large body of evidence, is that in which two (or more) sequential enzymes in a pathway transfer their common intermediate (the product of the first and the substrate of the second) directly, i.e. without releasing it to the bulk solvent. The evidence for this phenomenon, known as metabolic channelling, has been gathered by several methods, from X-ray crystallography to kinetic experiments in vitro.

One obvious situation in which metabolic channelling can easily occur is that of the "static" multienzyme clusters, complexes of several polypeptides with different enzymatic activities. There are many well-known examples of such clusters, including the pyruvate dehydrogenase and fatty acid synthase complexes and the ribosome. In such cases it is common knowledge that intermediates of the reactions catalysed by these clusters are not released to the solvent. Of particular interest is the case of tryptophan synthase, whose structure has been determined by X-ray crystallography to 2.5 Å resolution.

This cluster is an $\alpha_2\beta_2$ tetramer; the α chains convert indole 3-glycerol-phosphate to indole and D-glyceraldehyde 3-phosphate, and the β chains convert indole and L-serine to L-tryptophan. The images reconstructed by Hyde and co-workers show a tunnel going from the active centre of one α polypeptide chain to the active centre of one β chain (there are two of these tunnels per cluster).

It is through this tunnel that indole undergoes "facilitated diffusion" from one active centre, and is thus not released to the solvent. One obvious advantage of this mechanism is that if indole were to be released to solution it would be able, due to its hydrophobic character, to diffuse freely across the plasma membrane and thus escape. Only after site-directed mutagenesis of the tryptophan synthase, to cause indole to leak from its channel, were satisfactory yields achieved.

The channelling of metabolic intermediates, here known as static channelling or channelling via static complexes, is not, however, restricted to stable multienzyme complexes; it can also take place via short-lived enzyme complexes referred to as dynamic channels. Channelling between dynamic complexes of enzymes is in some ways harder to demonstrate since the enzyme-enzyme complexes are not stable and are thus not isolatable. Of course this classification describes two extreme situations, and one may consider that in reality enzymatic complexes have properties that are a continuum between the two extremes.

There is a remarkable correlation between the organizational state and the functionality of supramolecular enzyme complexes: static complexes are found in anabolic pathways such as amino acid, nucleotide and protein biosynthesis. In these pathways most of the intermediates have no other function or destination in the cell. In some cases, as for indole in the tryptophan synthase reactions in wild-type strains, there is essentially no leakage of the metabolite to the solvent (perfect channelling).

Dynamic complexes occur in amphibolic pathways like glycolysis and the Krebs tricarboxylic acid cycle (TCA) cycle where there are various flow-bifurcations. In this case there are metabolites at branch-points, which are known (e.g. from NMR measurements) to exist in soluble pools. However this does not rule out the possibility that a proportion of these substances is bound to enzymes (either individually or in complexes); in some cases, e.g. for oxaloacetate in mitochondria, the number of binding sites exceeds the number of molecules extractable using perchloric acid.

Indeed there is evidence that the role of enzymes is not simply catalytic but also structural. Based on the high concentration of enzymes and the relatively low concentration of some of their intermediates in glycolysis, it has been argued that an important role of the cytoplasmic enzymes may be that of buffering the concentrations of their ligands. In this case enzyme-intermediate complexes may act as better substrates for the next enzymes in the pathway than are the soluble intermediates themselves.

In a series of elegant in vitro kinetic experiments, D.K. Srivastava and S.A. Bernhard observed channelling of NADH among some dehydrogenases. The basis of their experiments is the buffering of NADH by the first dehydrogenase, such that if channelling occurs then the rate of the reaction of the second dehydrogenase calculated from the concentration of free NADH present will be different from the experimental rate. A requirement for these experiments is that the first dehydrogenase must be in large molar excess with respect to NADH which is in turn in molar excess over the second dehydrogenase; also the dissociation constant of the NADH from the first dehydrogenase must be known.

Srivastava and Bernhard also showed a remarkable correlation between the stereospecificity of NADH binding to dehydrogenases and the channelling of NADH between pairs of dehydrogenases. It turns out that in order for channelling to ocurr the two dehydrogenases must be of opposite chiral specificity to NADH. P.B Chock and coworkers reported inconsistencies of this analysis, based in part on their own repetition of these experiments, but it is now clear that this was due to the fact that they failed to reproduce key features of the original experiments, such as the ionic strength. The remarkable correlation between enhanced catalysis and opposite chiral specificity of the dehydrogenases (and no enhancement for equal specificities) is the major factor supporting the channelling model and this was never addressed by Chock's group.

There are a number of other cases of experimental results that point to channelling in dynamic complexes, but which have not universally been accepted. J. Ovαdi and T. Keleti proposed channelling of glyceraldehyde-3-phosphate from aldolase to glyceraldehyde-3-phosphate dehydrogenase but Kvassman et al., although reproducing Ovádi & Keleti 's results, argued that these were also consistent with a free diffusion mechanism. B. Vértessy & Ovádi showed data that suggested the channelling of dihydroxyacetone phosphate from aldolase to glycerol-3-phosphate dehydrogenase, although Chock & Gutfreund offered an alternative interpretation.

J. Weber & Bernhard suggested direct transfer of 1,3-diphosphoglycerate between glyceraldehyde-3-phosphate dehydrogenase and 3-phosphoglycerate kinase; however J. Kvassman & G. Pettersson argued that if autocatalytic effects of NAD+ were taken into account then channelling it would not be necessary to invoke. Unfortunately in most of these reports the data were not obtained under the same conditions by the different groups; there has also been a certain tendency amongst those who argue against channelling to do so despite the fact that only data which also support free diffusion are shown. There are, however, cases of metabolic channelling which occur through transient enzyme complexes and whose existence is unchallenged, despite the fact that, as in the previous cases, the actual multienzyme complex has never been detected. One example is that of electron transfers between enzyme-bound flavin coenzymes.

The Krebs tricarboxylic-acid cycle (TCA), has been extensivly studied in terms of its supramolecular organization. P.A. Srere has developed the concept of the metabolon, a complex of the enzymes of the TCA cycle which is anchored to the inner mitochondrial membrane. The metabolon can be obtained intact from sonicated mitochondria and there is evidence that the intermediates of the TCA cycle are channelled in the metabolon. Data that support this view were obtained in situ with gently disrupted (sonicated) mitochondria and in vivo in yeast cells using nuclear magnetic resonance techniques with-propionate as the substrate.

Using chromatographic techniques, Gontero et al. isolated a complex of 5 enzymes of the Calvin cycle from spinach leaves while Persson and Johansson isolated a 6-enzyme complex from the same source, using partition in an aqueous 2 phase-system. Interestingly enough, no interaction between purified enzymes of these complexes could be demonstrated in vitro. This strengthens the argument that one should study viable cells, and that if for technical reasons this is not possible that one should try to emulate the relevant physiological environment as much as possible.

The words of Henrik Kacser could not be more appropriate: "But one thing is certain: to understand the whole, you must look at the whole". One of the reasons which is arguably dominant in determining the difficulty in demonstrating dynamic complexes in vitro is the possibility that, in the intact cell, these complexes might be stabilized by other cellular structures. The MTL is an attractive candidate for such interactions but membranes can also play an important role.

The interaction of muscle cell glycolytic enzymes with actin is a remarkable example but other specialized cell structures are certainly also involved. It may also be that enzyme-enzyme complexes can affect the strength of allosteric effects (or even create new ones). It is also worth pointing out that the inability of a cloned enzyme to affect the flux through a pathway may be due either to the fact that they have a low flux-control coefficient or because they are unable to participate in channelling due to their expression at unsuitable concentrations or in an unsuitable location; to hope to distinguish these one should clone the relevant genes down as well as up.

For pathways such as glycolysis there is a distinct structure-function duality. This is evinced for example in the reversible, "ambiquitous" partitioning of glycolytic enzymes between cytomatrix-bound and aqueous cytoplasm-free forms. The direct-transfer "channelling" scheme is probably immaterial for the unbound enzyme fraction in the bulk cytoplasmic solution. Depending on the cell type the "whole cell" concentration of the individual glycolytic enzymes is in the range 1-100 μM. Considering the high, bulk "viscosity" of the cytoplasm relative to the translational motion of the macromolecles, it is unlikely that the binary complex, consisting of enzyme1 bound to its nascent product molecule, would last long enough to form the requisite ternary complex with enzyme.

However, for the cytomatrix-associated enzyme population, the microenvironmental situation is quite different. The "local" enzyme concentration is probably much higher than the averaged, "wholecell" value. In addition, the total macromolecular density at the cytomatrix-"cytosol" interfaces (=100Å into the bulk phase) is extremely high, probably similar to that within the mitochondrial matrix. Recent in situ measurements of the mitochondrial matrix bulk viscosity show the diffusion coefficient for small-metabolite-sized particles to be as much as 30-fold smaller than that for normal water.

With such high "local" enzyme concentrations, along with the high microenvironmental viscosity and high activity coefficients for protein-based reactions in such a "crowded" medium, the efficacy of formation of channelling ternary complexes has been calculated to be greatly enhanced. The size of the free metabolite pools for the dynamically interacting (pairwise) enzyme systems will also depend on molar ratios of the metabolically-consecutive enzymes, as well as on the respective enzyme kinetic mechanisms; moreover, the physiological grounds for channelling may not always entail a kinetic flux advantage for the multienzyme system.

Meanwhile, one should be cognisant of the weight of the in vivo evidence for channelling in these pathways, obtained from the many whole-cell studies, as well as the correlation with in vitro indications of a "physiological" regulation of enzymecytomatrix associations and of the control of enzyme-enzyme interactions at pathway branchpoints. The ambiquitous character of the enzyme organisation - and more specifically the variability in the degree (and spatial locale) of channelling—in the branching amphibolic pathways may reflect the need to maintain adequate catabolite pools for proper responsiveness of such multifunctional processes to varying cellular demands. It is within the localised cytomatrix microenvironments that channelling of amphibolic flow may therefore be expected.

Modelling Strategies for Enzymology Invivo

All interpretations of experimental results are based on models, and metabolic channelling is no exception. Generally, enzyme models exhibit non-linear behaviour, and it is very difficult (and dangerous) to make predictions from those models by simple reasoning alone. For that purpose one has to formulate the model in mathematical terms and use the equations to calculate the values that the variables in question will take as a function of different starting conditions (parameters). These calculations are best carried out using computer programmes to minimize errors and to process the calculations at an acceptable speed.

P.Smolen and J. Keizer simulated a model of dynamic channelling with the direct transfer of NADH between dehydrogenases in mind. This group concluded from their simulations that the conditions for the ternary complex dehydrogenase1-NADH-dehydrogenase$_2$ to form are in accordance with the estimates of K_m for the oxidation of the complex dehydrogenase1-NADH by dehydrogenase$_2$. More recently they have extended their analysis to complexes of three dehydrogenases which would cycle the redox couple NAD^+/NADH. They concluded that there would be novel methods for regulation of the redox state of the couple which are consequent upon the channelling mechanism.

One consequence of metabolic channelling that has frequently been mentioned is that the operation of the catalytic path through the channel would be expected to reduce the size of the soluble pool of the intermediate compared to the case where there the reaction proceeded exclusively via diffusion in the solvent.

Cornish-Bowden showed that this is not always the case by simulating a model of metabolite channelling through a dynamic bi-enzyme complex. He then argued that this reduction of soluble pool size could never happen with a dynamic channel, but it was later shown that this generalization does not hold: channelling can decrease the pool size substantially. It was also shown that if channelling operated through a static bi-enzyme complex then the size of the soluble pool could be decreased to an arbitrarily low level by increasing the proportion of flux through the channel.

S.P.J. Brooks and K.B. Storey investigated the possibility of the existence of a complex of glycolytic enzymes in muscle cells. They used in vitro data for the association constants of the several complexes (enzyme-enzyme, enzyme-F-actin and enzyme-calmodulin) and relative activities of these complexes to conclude that complexes of enzymes on F-actin "may not exist"; complexes of some sequential glycolytic enzyme pairs could exist to a significant degree and increase the glycolytic flux, and the binding of phosphofructokinase with F-actin could be a regulatory mechanism to control glycolytic rate.

This is far from the reality of muscle cells where the glycolytic enzymes are responsible for very high net fluxes. Also not considered in this analysis was the effect of the intermediate metabolites, which is very important, as shown for example in the studies of Smolen and Keizer—the association constant for an E1-E2 enzyme complex is certain to be different from that when the intermediate is part of the complex (E1-M-E2).

H.M. Sauro and H. Kacser, from a theoretical metabolic control analysis of a model of static channelling, predicted that the increase of the logarithmic flux of a binary-enzyme complex when the logarithmic concentration of both enzymes is increased simultaneously is not linear. Many of the supposed consequences of channelling through multienzyme complexes are specifically dependent on this feature. Perhaps the most frequently cited consequence of channelling is the reduction of the transient time of a reaction sequence, numerically equal to the sum of the metabolite concentrations divided by the pathway flux. A lower bound on the value of the transient time, τ, for the enzyme reaction (viewed as an intermediary metabolic process) can, under simplified (viz. pseudo-first order) steady-state conditions, be expressed as follows:

$$\tau = \frac{1}{\Omega \cdot D_{ES} \cdot R_{ES} \cdot f_e \cdot f_g \cdot [E]_T \cdot N_0} + \frac{k_{-s} + k_{cat}}{k_{+s} \cdot k_{cat} \cdot [E]_T}$$
$$= \tau_d + \tau_r$$

where Ω is a steric-orientation factor (e.g. a geometric solid-angle within the range $0 < \Omega = 4\pi$) relating to the approach of the substrate to the enzyme active site D_{ES} is the sum of the diffusion coefficients of E and S (where usually $D_{ES} \approx D_S$), R_{ES} is the "reaction distance" (viz. the sum of the radii of the substrate molecule and of the "recognition volume" of the enzyme active site); fe is an electrostatic term arising from the possible influence of the net charge of the globular protein on an approaching, charged substrate molecule (with typical values of f_e ranging from 0.1 to 10, depending on whether the algebraic product of the net charge of the globular protein and that of the substrate molecule is positive or negative, respectively); f_g is a "gating" term due to the potential effect of local motions of proteinaceous lobes surrounding the active site (where typically $f_g = 0.5$ if slow protein-dynamical gating is identifiable);$[E]_T$ is the total enzyme concentration;. N_0 is Avogadro's number; k_{+s}, k_{-s} and kcat are the (intrinsic) unitary rate constants for the binding of substrate to the enzyme, the release of substrate, and catalytic turnover respectively.

Such effects can only be investigated if the spatial dimension is explicitly part of the model. Reaction-diffusion models are unfortunately more difficult to simulate than normal kinetic models and there are no software packages available for this purpose.

Chapter 8

Enzyme Genetic Programming

Most genetic programming systems represent the programmes they are evolving using either explicit or indirect context. In biological representations, shape is an innate form of implicit context. The programme representation used by enzyme genetic programming is modelled upon the representations used by biology to represent enzyme systems, and models shape as a pattern which describes the role of each component defined within the representation.

In genetic programming, we are interested in evolving programmes. Enzyme GP models computation as a network of interacting functional elements, input nodes and output nodes. This model is general enough to capture most of the programmes that are evolved by other GP systems, including the tree-structures evolved by standard GP. The role of each of the nodes within the network can be defined by the nature of its outputs. For a functional element, this is the result of applying its function to the data supplied at its inputs i.e. it is determined by both its function and its inputs. Its inputs, more specifically its input connections from other network nodes, determine the context within which its function is applied.

Use of Expicit Context

The most common representations used by GP declare explicit

connections between components. Consequently, a node's context is also recorded explicitly. Examples of this form of representation include the parse tree of conventional genetic programming and linear GP representations. This use of explicit context has a number of implications for these representations when subjected to evolution. First, a node's context is typically determined by its position within the representation.

Crossover operators (unless designed to be context-preserving and therefore more computationally intensive) tend not to preserve this positional information and, consequently, tend to disrupt context. The result of this is that crossover is more likely to cause macro-mutation rather than meaningful recombination. Second, where explicit context is used there is usually a one-to-one mapping between representation and programme, such that changes to the representation lead directly to changes within the programme. The variation operators therefore act directly upon the programme, implying that evolution is determined solely by processes of variation and selection. The representation, by comparison, has no dynamic role to play within the evolutionary process.

Indirect Context

In some other GP representations, connections between components are specified using indirection. Typically each component is assigned a reference (which may be a location, a number or an arbitrary code) and other components specify their input connections using these references. An interesting facet of representations that use indirect context is that a component only becomes active in the programme if another component expresses a connection to it. Otherwise it is recessive. This has some important implications. If a new component is added to the representation during recombination, it will only become active in the programme if its reference is addressed by an existing component or it is an output node. This effect is termed variation filtering, since the representation only expresses certain variation events in the programme whilst filtering out others.

The trouble with most representations that use indirect context is that components are assigned references arbitrarily. Usually there is no correlation between component reference and component behaviour within, and more signifi-cantly, between programmes. Accordingly, components with different behaviours can have the same reference and components with the same behaviour can have different references. This implies that the meaning of a component's input context depends upon which other components are present in its environment.

If this environment changes following recombination, this will cause the component's context—and therefore its role within the programme—to change. Hence, indirect context does not maintain its meaning following recombination. However, it is conceivable that the population might be able to evolve a correlation between component reference and component behaviour over the course of time, especially as the population becomes more homogenous. In this sense, the meaning of indirect context is evolvable. Representations that use indirect context also have a number of other interesting evolutionary behaviours.

For instance, mutation is able to change the structure of the programme by targeting the references that specify connections. Furthermore, mutation or recombination can lead to recessive components becoming re-activated. This may encourage back-tracking behaviours during evolutionary search. Also, recessive components are still exposed to evolution: encouraging useful forms of neutral evolution.

Implicit Context

Implicit context is an innate property of biological representations and, as we shall see, the programme representation used by enzyme GP. The principle behind implicit context is that a component's input context is defined in terms of behavioural properties of the component it would like to receive input from. For example, the behaviour of a bio-chemical is dependent upon its physical shape and chemical properties. Enzymes, which conceptually receive their input in the form of bio-chemicals, express their preference for these inputs by the shape and chemical properties of their binding sites. Since the bio-chemicals that they bind have physical shape and chemical properties complementary to these binding sites, the binding sites are implicitly describing the context of the substrates they expect to bind.

Implicit context within programme representations captures many of the rich evolutionary behaviours possible with indirect context, but—because context is defined in a manner which is independent of any particular environment - does not suffer the same problems with loss of meaning following recombination. In particular, representations with implicit context also experience variation filtering. However, unlike with indirect context representations where loss of meaning makes this process arbitrary, variational filtering in representations with implicit context may be a significant source of evolvability: filtering out inappropriate variation whilst preserving meaningful variation.

If a new component is added to the representation (during recombination), it will only become active in the programme if it either fulfills an existing component's preferred context better than any existing component or it is an output node. If it does not fit into the context of the programme, it will become recessive. If an active component is removed from a programme (during recombination), then the components which received input from it will receive input from components that have functional properties most similar to the one that was removed -which may be components that are currently recessive.

If an existing active component is modified (by mutation), it may become recessive if the context it provides no longer fits with the rest of the programme. If a recessive component is modified, it may become active if the context it provides now fits better into the programme than the context provided by an active component. In effect, the combination of implicit context and variation filtering could result in a tendency to preserve existing contexts within the programme following the application of variation operators.

In turn, this will tend to preserve the existing behaviour at the programme outputs (which may come from more than one programme following recombination). This idea is illustrated in figure 1. However, there are certain pre-conditions for implicit context being useful in enabling meaningful recombination and variation filtering. Precision is the generality of the implied context: the degree to which it suggests an actual context. If implicit context is too general, then the behaviour of the programme will not be obvious from the representation, making the mapping between representation and programme unstable and therefore easily disrupted by the application of variation operators. In turn, this will make meaningful recombination unlikely.

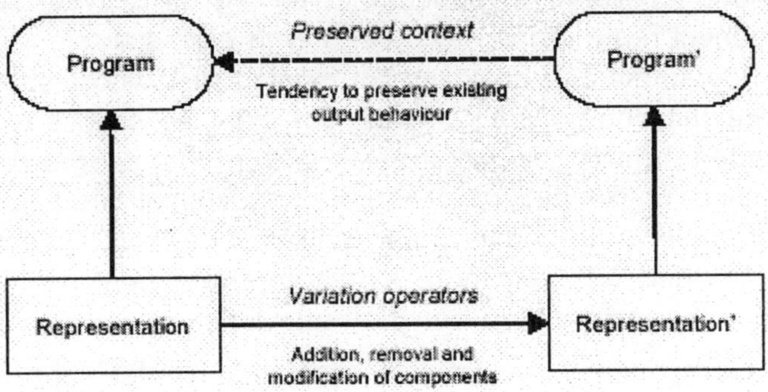

Figure 1: The combination of implicit context and variation filtering causes a tendency to preserve existing context and programme output behaviour.

Related to precision is the issue of specificity: how well an implied context is able to identify an actual context. If a component's implied context is too unspecific, it is likely to form different actual contexts within different programmes, meaning that variation filtering will tend to carry out arbitrary behaviours. Finally, there is the issue of accuracy: the ability of implicit context to match actual context given the availability of components within the representation and any constraints upon connections between components that are placed upon the behaviour of the programme.

Given that these limitations are unavoidable, there is little that implicit context can do to overcome them. Nevertheless, it is important that when the preferred context is not available, the implied context should match the nearest available actual context. In turn, this requires that a distance metric can be defined between implied and actual contexts, such that the greater the distance between contexts, the greater is the difference in component behaviour.

Enzyme Genetic Programming and Implicit Context

In enzyme GP, programme components are modelled upon biological enzymes. In addition to having an activity, each component has a shape which declares its expected role within the programme and analogues of binding sites whose shapes declare the component's implicit context—the shapes of its expected inputs (or substrates) within the programme. However, unlike biological enzymes, individual programme components carry out only a limited range of functional activities and typically more than one component within a programme would carry out the same function.

Consequently, for implicit context to confer suf-ficient precision and specificity, it would be insufficient for shape to capture only the component's activity. Accordingly, shape captures the entire form of the component: both its activity and the shapes of its binding sites. Likewise, binding sites describe the complete shape of the components they expect to bind.This has an interesting implication. Since a component's shape captures the shapes of its binding sites, and these shapes capture the entire form of its expected substrates, it is also capturing the activity and binding site shapes of these expected substrates: and recursively, the shapes—and therefore the activities—of all the components in the programme fragments that develop below its inputs.

This is not a behaviour that occurs in biological enzyme systems, since enzymes interact through chemical intermediaries and even then do not capture the entire shape of their substrates. Nevertheless, it would appear to be a useful mechanism in increasing the precision and specificity of implicit context within enzyme GP. The shape of a programme component is defined using a measure called functionality: a vector which describes the component's position within an activity reference space. This reference space has one dimension of unit length for each member of the GP function and terminal sets. The functionality, *F*, of a programme component is defined as follows:

$$F(component) = (1 - k) \, . \, F(activity) + k \, . \, F(binding\ sites) \quad (1)$$

where *k* is a constant that biases the functionality towards either the component's activity or the component's binding sites; *F*(*activity*), the functionality of the component's activity, is a unit vector situated in the dimension corresponding to the enzyme's function; and *F*(*binding sites*), the functionality declared by the component's binding sites, is defined:

$$F(binding\_sites) = \frac{\sum_{i=1}^{n} F(site_i) \cdot strength(site_i)}{\sum_{i=1}^{n} strength(site_i)} \quad (2)$$

i.e. the average of the functionalities corresponding to its binding sites weighted by the strength (a number between 0 and 1) of each binding site. In effect, a component's functionality declares an expected activity profile of the components that occur in the programme fragment of which it is the root, weighted by depth and biased by the strength of binding sites. Functionality space itself is continuous and the distance between functionalities is calculated using vector subtract on. This reflects the difference between their activity profiles, meeting the requirement that the greater the distance between contexts, the greater is the difference in component behaviour.

Mapping Programme

A programme representation is the enzyme GP analogy of a biological genome: stored as a linear array of programme components sub-divided into input terminals, functional elements and output terminals. In addition to an activity declaration (a member of the function or terminal set), each component that processes inputs, i.e. functional elements and output terminals, contains a list of potential binding sites, each of which has a functionality and a strength. The objective of mapping programme representation to programme is that each active input of each active component will be connected to the output of the component for whose shape its corresponding binding site has the highest specificity i.e. the shortest distance between functionalities.

A number of different mapping processes are possible. The mapping process begins with expression of the output terminals, which then choose substrates whose shapes are most similar to the shapes of their strongest binding sites. These substrates are now considered expressed and, if they require inputs, attempt to satisfy them by binding their own substrates. This process continues in hierarchical fashion until all expressed programme components have satisfied all of their inputs. Each output terminal is expressed exactly once, whilst input terminals and functional elements can be expressed once or not at all.

Where more than one component chooses the same substrate, the output of the substrate is shared between them.

The programmes are required to have strict tree-structures. This constraint is handled within the mapping process by checking for the presence of cycles before a new connection is made. If a connection to a substrate would result in a cycle, then an alternative (less preferred) connection must be made. Unfortunately this cycle checking brings the time complexity of the mapping process above linear, but is an unavoidable overhead of developing a cyclic graph into a tree structure. An alternative mapping process, based upon transforming the programme representation into a network random key representation, has also been used within enzyme GP, but has a higher average-case complexity than the process outlined above.

Evolution of programme representations occurs within a spatially-distributed parallel genetic algorithm. Bounded-size programme representations are constructed randomly to fill the initial population. New programme representations are created through crossover and mutation of existing programme representations.

Crossover in enzyme GP is somewhat different to sub-tree crossover in conventional GP and takes advantage of the fact that added components need not replace existing components. In sub-tree crossover, one sub-tree is always replaced by another. In enzyme GP crossover, a contiguous group of components is copied from one solution to another without removing any existing components (with the exception of output terminals, whose numbers remain constant).

This transfer operation is complemented by a converse operation which removes contiguous groups of components from solutions. Each of these operations is used, non-deterministically, for 50% of crossover events. The effect of the remove operation is to balance solution sizes so that recombination has an overall neutral effect upon solution size within a population. For both transfer and remove operations, the number of components targeted is chosen randomly within an upper limit (t_u). Mutation targets both functionalities (m_f) and activities (m_a).

Representation Properties

The comparative analysis of the performance of enzyme GP upon a range of problems in combinational logic design. These have shown that the approach is able to compete favourably against indirect context representations upon most problems that it has been applied to. A more direct comparison is shown in figure 2, where fitness curves for enzyme GP with functionality shapes are shown against fitness curves for enzyme GP with randomly generated shapes. These random shapes effectively specify an indirect context, where there is no relationship between pattern and behaviour. It can be seen that the fitness curves for enzyme GP with random shapes initially grow at a high rate, but fall to a relatively low rate once they reach a certain fitness level.

This suggests a high level of disruptive variation, which initially benefits search but later becomes a hindrance to effective exploitation. The curves for enzyme GP with functionality shapes, by comparison, demonstrate a steady rate of decay in fitness growth, indicating a more structured search which continues until the optimum is found. Performance statistics for enzyme GP with both types of shape are listed in table 1; and indicate that performance is considerably better with functionality shape than with random shape. This supports the notion that implicit context captures more meaningful context than indirect context.

Recombination and Context preservation

Table 1 also compares the performance of enzyme GP both with and without recombination. When functionality shapes are used, recombination leads to a significant improvement in performance. This occurs even though the problem's fitness

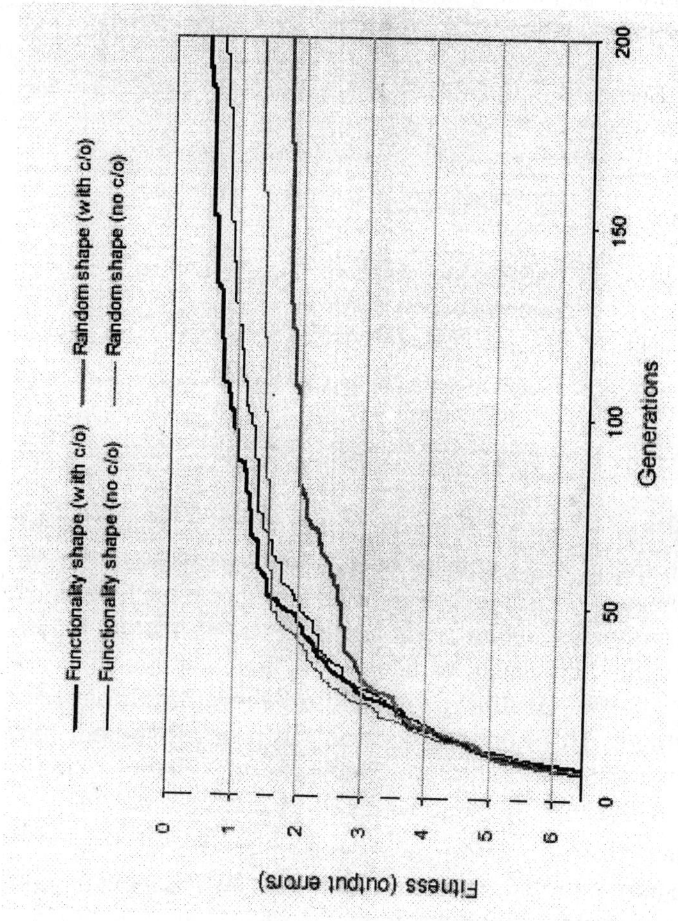

Figure 2: Comparing mean number of output bit errors when evolving two-bit multipliers

landscape is not particularly suited to recombination, and illustrates that functionality shapes are able to preserve context both within and between programme representations. When random shapes are used, however, recombination impairs performance: suggesting that recombination disrupts rather than preserves context.

Table 1: Success rate and average solution time (in generations) for evolving two-bit multipliers in a population of size 324 with k = 0.3, m_f = 2%, m_a = 1:5% and t_u = 5.

Shape	Variation operators	Success rate	Average
Functionality	Mutation and recombination	70%	98
Functionality	Mutation only	55%	79
Random	Mutation and recombination	26%	69
Random	Mutation only	30%	64

Figures 3 and 4 show examples of the behaviours that occur when transfer and remove operations are applied during recombination. The transfer operation cannot directly replace any components within a programme representation (other than outputs). However, as figure 3 illustrates, it can still lead to behaviours at the programme level that look like replacement. This happens because the new component has subsumed the role of the former component, offering a closer match to the input context declared by the parent node.

Nevertheless, the former component is still present within the programme representation and, if the new component was removed, could resume its former role in the programme. Figure 3 also shows how subsumption is able to preserve the behaviour of the programme below the component that has been replaced, indicating that the new component has a functionality similar to that of the former component, and illustrating the context preserving nature of recombination in enzyme GP.

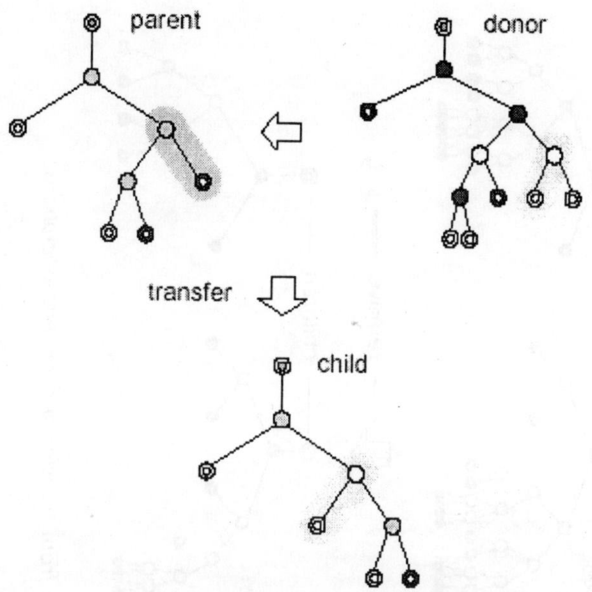

Figure 3: A simple example of subsumption resulting from a transfer operation.

Figure 4 shows two other behaviours which often occur during recombination: insertion and deletion. In the example on the left of the figure, a transfer operation leads to a new sub-tree being inserted between two existing components, something which is not possible using sub-tree swapping recombination in standard GP. Again, the existing sub-tree below the root of the inserted sub-tree is preserved, and from the perspective of the component (the output terminal) above the inserted sub-tree, the new context is related to the former context: and presumably a closer fit to the implicit context declared by its binding site.

In fact, insertion is a special case of subsumption where the subsuming component happens to declare an input context

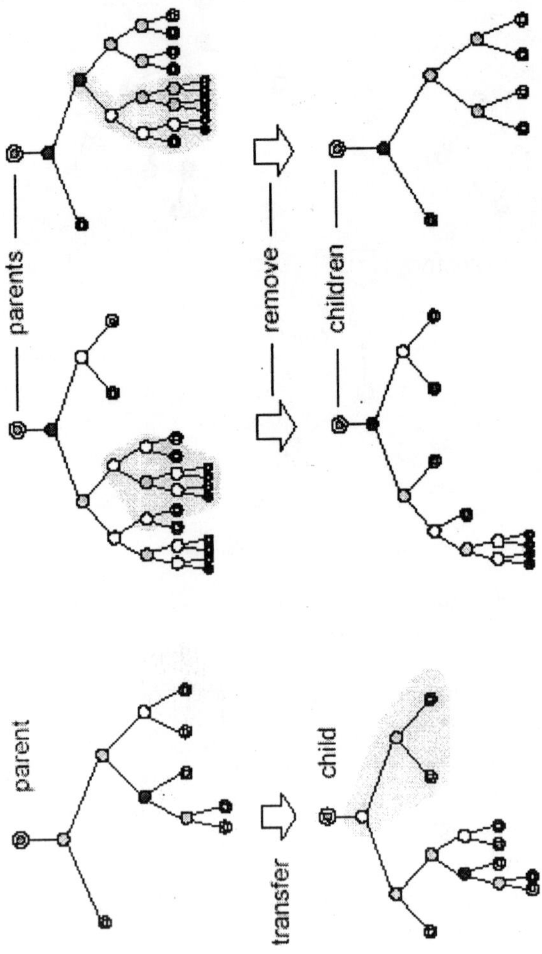

Figure 4: Behaviours resulting from transfer and remove operations

which matches the role of the component that it subsumed. In the other two example in figure 4, removal operations lead to parts of programmes being deleted. In the example in the centre, two sub-trees are affected by a single removal operation. This shows how recombination operators in enzyme GP operate upon groups of components at the representation level rather than structures (in this case sub-trees) at the programme level: a behaviour which is presumably more disruptive, but also more expressive, than conventional sub-tree crossover.

In the example on the right of the figure, an entire sub-tree is removed from between two nodes. This is the complement of the behaviour which occurs in the transfer example on the left of the figure, and shows how a subsumption operation might become undone. In both of these removal examples, the programme below the deletion remains unaffected, showing how the remove operation also encourages context preservation.

Filtering Variation

Conceptually, all the components defined in a representation occupy some position in a subsumption hierarchy. At the top are the expressed components: those which appear in the programme. Below these are any redundant copies of the expressed components. Below these are components that have been subsumed by the expressed components; and further down, components that were subsumed by components further up the hierarchy which have since themselves also been subsumed.

At the bottom are components which have never been expressed but which could in principle be expressed if all the components above them in the subsumption hierarchy were removed. All the behaviours that occur in enzyme GP are a result of variation operators modifying this subsumption hierarchy: either by adding new entries, by removing entries or by re-ordering existing entries (which is what mutation operators are essentially doing). Nevertheless, the operators are not aware of this subsumption hierarchy. They blindly add, remove and re-order entries—only causing change in the programme when they happen to add, remove or re-order entries at the top of the hierarchy. All other changes are absorbed into the lower echelons of the hierarchy.

Subsumption supports the idea that the programme representation used by enzyme GP promotes meaningful variation filtering: since it is clearly the way in which the representation describes the programme, rather than the action of the variation operators, that allows subsumptive behaviours to occur. Further evidence for the existence of variation filtering can be seen in the relationship between patterns of growth in the programme and in the representation.

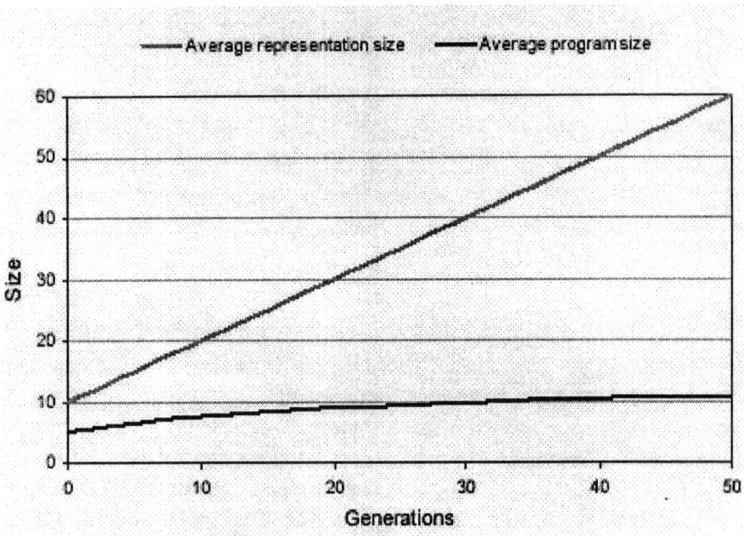

Figure 5: Comparing representation size growth against forced program size growth in a population of size 100 with no selection pressure, no removal operation and no mutation.

Figure 5, by comparison, compares programme and representation size growth when the population is not undergoing selection and only the insertion operator is used. Consequently, any problem-specific programme size influence is removed. Again, programme size growth is considerably lower than representation size growth, implying that the proportion of un-expressed components in a programme representation, on average, increases over time. This effect can be accounted for by a process of variation filtering, where the un-expressed portion of the representation contains those components which have been inserted into the representation but not propagated to the programme.

This un-expressed portion of the programme representation constitutes the recessive part of the subsumption hierarchy described above. Although it would not normally grow at the rate seen in this example, it is interesting to consider whether it could have a function other than filtering out components that do not fit into the context of the programme. One such function is evolutionary back-tracking.

If the population were to evolve towards a local optimum, these recessive subsumption hierarchies contain information that could be used as a means of escape to some previous point before the population converged upon the local optimum. Other functions concern behavioural plasticity: the possibility of recessive components becoming expressed during operation to compensate for failure of expressed components or changes in the programme's requirements.

Chapter 9

Computational Design of Enzyme-like Proteins

Protein design is the generation of proteins with novel functions, including the catalytic rate enhancement of chemical reactions at which natural enzymes are so efficient. The ability to design an enzyme to perform a given chemical reaction has considerable practical application for industry and medicine, particularly for the synthesis of pharmaceuticals. Significant progress has been made at enhancing the catalytic properties of existing enzymes through directed evolution.

In contrast, the design of proteins with novel catalytic properties has met with relatively limited success. This chapter present a research report of Daniel N. Bolon. He gives a general computational approach for the design of enzymelike proteins with novel catalytic activities and details as follows.

The use of transition state analogs as haptens to elicit catalytic antibodies has been the most successful technique to date for generating novel protein catalysts. Natural enzymes combine transition state stabilization with precisely oriented catalytic side chains. While a reactive hapten has been used in the generation of an antibody with a powerful nucleophile at the active site, current catalytic antibody technology does not efficiently select for both catalytic side chains and tight non-covalent affinity in the same molecule.

The relationship between the general backbone fold of an enzyme and its catalytic properties is not well understood. This is particularly relevant to catalytic antibodies that are currently constrained to the antibody fold and which have yet to show catalytic activity on par with natural enzymes. Rational, non-computational design efforts have recently succeeded at altering the catalytic reactions of two different enzymes. Cyclophilin, a cis-trans isomerase of XPro peptide bonds was engineered into an endopeptidase by grafting a triad of catalytic residues commonly found in serine proteases at the binding cleft.

In a more complicated design effort, A. R. Fersht and co-workerswere able to convert indoleglycerol-phosphate synthase into phosphoribosylanthranilate isomerase. The naturally occurring versions of these enzymes share a similar fold and catalyze consecutive reactions in the tryptophan biosynthetic pathway with the product of the isomerase being the substrate for the synthase. Based on a comparison of the crystal structures of both natural enzymes, loops were altered in the synthase to resemble the target isomerase, followed by directed evolution. These two successful design efforts both utilized starting proteins with substantial substrate affinity and altered existing active site residues in order to catalyze new reactions.

Compared to altering substrate specificity or catalytic mechanism, one of the most significant challenges of enzyme design utilizes a starting protein scaffold devoid of substrate binding and catalytic activities and introduces residues that convert the scaffold into an enzyme. Computational techniques have been used to design novel metal binding sites into proteins. By leaving one of the primary coordination spheres of the metal un-ligated by the protein, nascent metalloenzymes with a variety of oxygen redox chemistries have been generated. The general design of metalloenzymes to react specifically with more complicated organic molecules has not been demonstrated.

Numerous studies of natural proteins have revealed the basic principles of enzymatic catalysis, including proximity and orientation of substrate molecules, transition state stabilization, acid-base catalysis, and covalent catalysis. The continued improvement in computer speed combined with the development and improvement of computational protein design algorithmsand forcefieldsmakes this an opportune time to apply his understanding of enzyme mechanism to the rational, computational design of enzymes with novel functions.

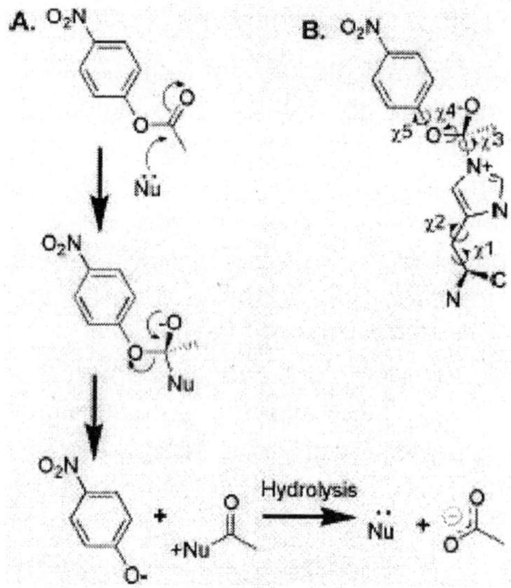

Figure 1: A) Nucleophile mediated catalysis of PNPA hydrolysis. B) High-energy state structure used in the computational active site scan

He model a high-energy state along a specified reaction pathway including the interaction of at least one catalytic residue with the highenergy state. The computational techniques described below, which have been incorporated into the ORBIT (Optimization of Rotamers By Iterative Techniques) protein design software, identify catalytic sites within a protein structure as well as mutations necessary to accommodate substrate binding (that is, to build the active site). The histidine mediated nucleophilic hydrolysis of p-nitrophenyl acetate (PNPA) into p-nitrophenol (PNP) and acetate (Figure 1A) was selected as the target reaction.

This reaction has a moderate activation barrier in pH neutral aqueous solution and is similar to the ester and carbonate hydrolysis reactions studied by the first catalytic antibodies. Nucleophile catalyzed hydrolysis is commonly achieved using a catalytic serine or cysteine in natural proteases. Histidine is often used as a general acid or base, but occasionally histidine acts as a nucleophile in phosphate transfer reactions. While histidine is not as powerful a nucleophile as activated serine or cysteine, it is sufficient to catalyze PNPA hydrolysis due to the moderate activation barrier of this reaction. The second order hydrolysis of esters by histidine side chains in a peptide based coiled coil system has been analyzed.

The peptides did not display substrate binding affinity, but the pKa and pH rate profile of reactivity towards PNPA could be modified by varying the local environment of the histidineand in one variant, hydrogen bonding to the transition state appeared to provide second order rate enhancement. For nucleophilic catalysis to be effective, the acylated enzyme intermediate must be unstable relative to substrate in buffer in order to increase the rate of product formation over the uncatalyzed buffer rate. An acyl-histidine intermediate would be less stable than either an acyl-serine or an acylcysteine intermediate. For these reasons, he modeled histidine as the nucleophile in his design.

Computational Materials and Methods

For the histidine-PNPA high-energy state rotamers, a backbone independent rotamer library was generated that included nucleophilic attack by both the Nδ and Nε atoms of histidine and attack on both enantiotopic faces of PNPA. The χ1 and

$\chi 2$ dihedral angles were based on histidine dihedral angles in a survey of protein structuresand were expanded by ± 1 standard deviation from the reported values.

Dihedral angles for $\chi 3$ (+/-30, +/-90, and +/-150 degrees), $\chi 4$ (+/-60, and 180 degrees), and $\chi 5$ (+/- 90 degrees) were based on canonical values for the relevant central atom hybridizations as described in the DREIDING force field. Bond lengths and angles as well as additional dihedral angles were optimized using the DREIDING force field.

All other side chains were modeled using a backbone dependent rotamer librarythat was expanded about $\chi 1$ and $\chi 2$ dihedral angles for aromatic residues, expanded about $\chi 1$ dihedral angles for aliphatic residues, and unexpanded for polar residues as previously described. The calculations used an energy function based on the DREIDING force field.

A Lennard-Jones 12-6 potential with van der Waals radii scaled by 0.9 was applied to all side chain/backbone and side chain/side chain interactions as previously described. An explicit hydrogen bond potential was applied to all groups that contained hydrogen bond donors and/or acceptors.

The 12-10 hydrogen bond potential used a well depth of 8. 0 kcal/mol, an equilibrium donor/acceptor distance of 2.8 Å, and a hybridization dependent angle term as previously described. Electrostatic energies were computed using a distance dependent dielectric constant of 40r with partial atomic charges as previously described.

Partial atomic charges for the high-energy state rotamers were generated using the same bond polarization scheme used for the other amino acids and included a net +1 charge on the histidine moiety and a net –1 charge on the high-energy state substrate (Table 1). A surface area based atomic solvation potential was selectively applied to the high-energy state rotamers in order to favor interaction of the substrate with the protein (i. e., only surface area burial of atoms on the high-energy state rotamers were considered). Surface areas for the solvation potential were calculated using the Connolly algorithmand the Lee and Richards definition of solvent accessible surface areawith unscaled atomic radii and an add-on radius of 1.4 Å. A hydrophobic burial benefit of 48 cal/mol/Åand a hydrophobic exposure penalty of 76.8 cal/mol/Å^2 were used with the two-body decomposition method of Street and Mayo. The solvation potential was not applied to polar atoms.

Table 1: Partial charges on computed high-energy state

Nucleophilic attack by Nε		Nucleophilic attack by Nδ	
Nδ1	-0.3	Nε2	-0.3
Hδ1	+0.5	Hε2	+0.5
Nε2	+0.6	Nδ1	+0.6
Cζ	+0.4	Cε3	+0.4
Oη1	-0.4	Oζ1	-0.4
Oη2	-0.9	+Oζ2	-0.9
Cι	+0.1	Cç	+0.1
Cμ	+0.2	Cë	+0.2
Nν	+0.6	Nμ	+0.6
Oo1	-0.4	Oν1	-0.4
Oo2	-0.4	Oν2	-0.4

The hydrophobic solvent accessible surface area of substrate atoms was used to evaluate recognition. For this assessment, solvent accessible surface area was computed using the Connolly algorithmand the Lee and Richards definition of solvent accessible surface areawith an add-on radius of 1.4 Å. The solvent accessible surface area of the substrate was determined using all atoms in the folded structure as well as using an unfolded reference that included only the high-energy state rotamer and the local backbone.

Active site designs were ranked by the fraction accessible surface area defined as the ratio of hydrophobic accessible surface area in the folded structure and the hydrophobic accessible surface area in the unfolded reference. Computed designs with zero total solvent accessible surface area for the substrates were considered inaccessible to substrate and were eliminated from further consideration.

Experimental Materials

The background mutation D26I was included in both PZD1 and PZD2. D26I was predicted by ORBIT in an independent calculation and results in increased thermodynamic stability similar to the previously reported D26A protein. Position 26 is distal in space to the designed active sites in both PZD1 and PZD2. The genes for PZD1 and PZD2 were constructed by site directed mutagenesis using the wild type thioredoxin gene (Invitrogen) cloned into PET-11A. Protein expression was induced with 0.5 mM IPTG from BL21(DE3) cells grown to mid log phase. Cells were lysed by sonication and pelleted twice at 20, 000g for 30 minutes.

The soluble fraction was brought to 60% acetonitrile, pelleted and rotory evaporated to half volume. For initial studies, purification was accomplished by reversed phase high performance liquid chromatography. For subsequent studies with PZD2, all protein samples were additionally purified by ion exchange and size exclusion chromatography to a purity of >99% as judged by silver stained PAGE. PZD2 was dialyzed extensively against 10 mM sodium phosphate buffer at pH 6.95. Kinetic experiments at 22°C were started by the addition of substrate dissolved in acetonitrile to buffer solution with and without PZD2 (final protein concentration of 4 μM).

Protein concentration was determined by UV absorbance in 6 M guanidinium hydrochloride assuming an extinction coefficient of 12400 M^{-1} cm^{-1} at 280 nm. Product concentration was determined by the change in absorbance at 400.5 nm assuming an extinction coefficient for deprotonated PNP of 19700 $M^{-1}cm^{-1}$. Final acetonitrile concentration was 1% for all experiments. The steady state rate of hydrolysis by PZD2 was corrected for the buffer rate. Burst phase hydrolysis assays were performed at a substrate concentration of 1.6 mM.

Protein concentration was 4 μM, with the exception of wild type thioredoxin. Wild type thioredoxin was assayed at 50 μM to improve signal to noise and extrapolated to a protein concentration of 4 μM. The dead time for this experiment was approximately 30 seconds.

The +42 mass unit species detected in the trapping experiment is a result of replacement of a hydrogen (-1) with an acetyl group (+43). 100 μM PZD2 in 10 mM TRIS at pH 7.0 was reacted with 1.6 mM PNPA to steady state conditions and a mass spectrum acquired. The same protein solution without PNPA was used as a control. Burst phase kinetics in this buffer system yielded essentially identical results as in phosphate buffer.

Computational Results

An important step in the design of a protein catalyst is locating the substrate relative to the protein scaffold. The location of binding sites for small molecule/protein docking is commonly evaluated by holding the protein fixed and performing a rotation/translation search of the small molecule using a grid based method. The design of a protein catalyst requires sequence changes in the protein scaffold and a grid based approach would require optimization of the protein sequence for each substrate position making it computationally intractable.

Instead he took advantage of the observation that the vast majority of natural enzymes contain one or more side chains in highly specific geometric relation to substrate. In the case of PNPA hydrolysis catalyzed by histidine, the histidine side chain should be positioned with a nitrogen atom approaching the carbonyl carbon adjacent to the scissile bond. In a procedure reminiscent of transition state analog synthesis for catalytic antibody design, he modeled a high-energy state of histidine catalyzed PNPA hydrolysis as a series of side chain rotamers (Fig.1B).

Modeling a high-energy state and optimizing the surrounding protein sequence for binding to the high-energy state may lead to a reduction in activation energy for the reaction and catalytic rate enhancement. Substrate affinity is explicitly included in the design because chemical similarity exists between the high-energy state and the substrate. The well-studied protein *E. coli* thioredoxinwas selected as a scaffold due to its favorable expression properties, thermodynamic stability, and successful history in computational design.

Naturally occurring thioredoxin can supply the reducing power for ribonucleotide reductase via its disulfide bond, but is essentially catalytically inert with respect to PNPA binding and hydrolysis. The high thermodynamic stability of thioredoxin suggests that it can tolerate destabilizing mutations that are required to build an active site. An active site scan was performed in order to identify favorable positions for the catalytic histidine as well as mutations necessary for substrate recognition and binding. In separate calculations using the ORBIT computational protein design programs, each position in the protein structure of thioredoxin was modeled using the set of high-energy state rotamers (Fig.1B). This strategy systematically limits the search to the relevant phase space where histidine and the substrate are properly positioned to undergo chemistry.

All other positions in the protein were allowed to chose (with proper consideration for rotamer flexibility) between their wild type identity and alanine in order to accommodate the substrate and to build the active site. After computing the optimal solution using algorithms based on the Dead-End Elimination theorem, positions that changed to alanine can be subsequently allowed to change identity to other amino acids in order to form better interactions with the high-energy state rotamer. The results reported below are based on the first stage calculation that only allows mutation to alanine.

The initial limited combinatorial complexity approach provides for tractable calculations and is expected to generate enzyme-like proteins that he call protozymes. As computer power and computational methods become more sophisticated, the need for a two-stage calculation will diminish. All 94 non-glycine, non-proline positions, a combinatorial complexity of approximately 10amino acid sequences that corresponded to 10rotamer sequences, were scanned in approximately two days on 14 195 MHz R10000 processors (Silicon Graphics) running in parallel.

Designs (resulting from the limited combinatorial complexity calculations) with the catalytic histidine at different positions in thioredoxin were ranked based on recognition of the high-energy state rotamer, using hydrophobic solvent accessible surface area of substrate atoms as a metric, as well as the simple requirement that the total solvent accessible surface area of substrate atoms be greater than zero in order to eliminate designs where the active site would not be accessible to solvent.

Experimental Result

The top two sequences from the limited combinatorial complexity active site scan were selected for experimental analysis. Protozyme design 1 (PZD1) contains two mutations required to introduce the catalytic histidine and to build the active site (F12H and Y70A) while PZD2 contains three mutations (F12A, L17H, and Y70A). Experimentally, both proteins demonstrated catalytic hydrolysis of PNPA at a rate significantly above background. Based on preliminary kinetic experiments, PZD2 shows the substrate atoms in a cleft created above the β sheet and between two α helices (Fig.2).

This cleft (Fig.3A) is not present in the wild type thioredoxin x-ray structure(Fig.3B), indicating that the space creating mutations (F12A and Y70A) are necessary to create the putative substrate-binding site (Fig.3C).

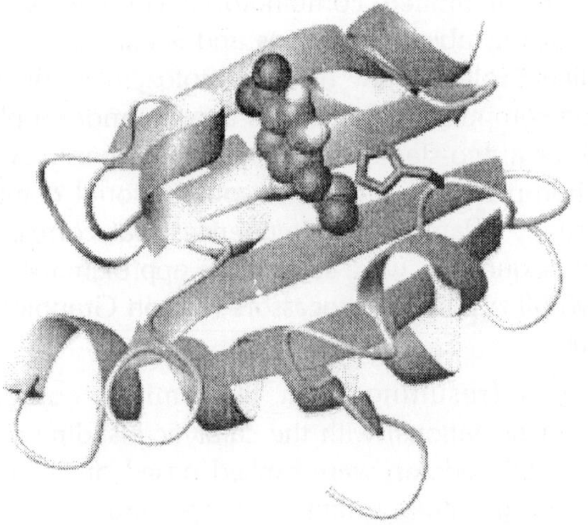

Figure 2: Computational design of PZD2

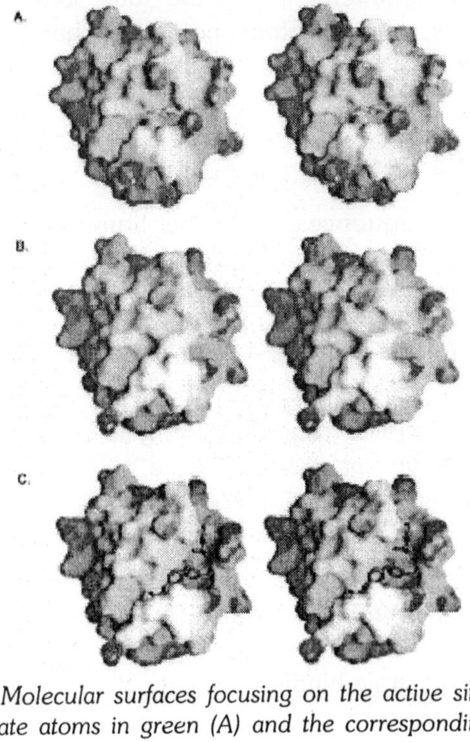

Figure 3: Molecular surfaces focusing on the active site of PZD2 with substrate atoms in green (A) and the corresponding region in the x-ray crystal structure43 of the wild type scaffold (B & C).

$$E + S \underset{}{\overset{K_s}{\rightleftharpoons}} ES \xrightarrow{k_2} P1 + EI \xrightarrow{k_3} E + P2$$

$$v = \frac{k_{cat}[E][S]}{K_m + [S]} \; ; \; k_{cat} = \frac{k_2 k_3}{k_2 + k_3} \; ; \; K_m = \frac{K_s k_3}{k_2 + k_3}$$

Figure 4: Kinetic model used to analyze the activity of PZD2.

The rate of PNPA hydrolysis by PZD2 at 22 buffer at pH 6.95 was experimentally determined over a range of substrate concentrations using standard Briggs-Haldane steady state treatment (Fig. 4). The reaction velocity of PZD2 demonstrated saturation kinetics with respect to increasing substrate concentration (Fig. 5) indicating that the molecule is acting as an enzyme-like protein with both substrate affinity and catalytic rate enhancement.

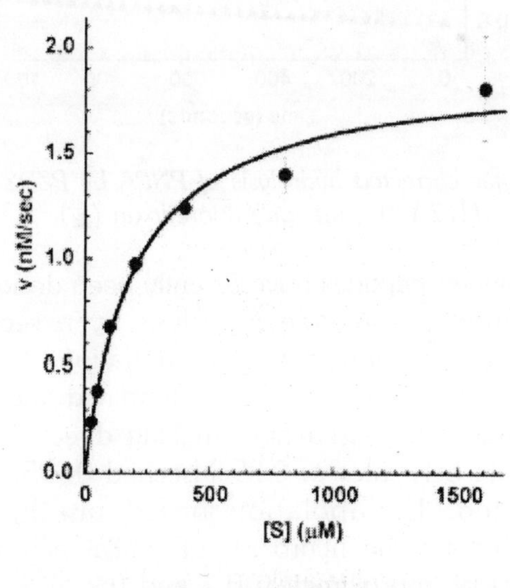

Figure 5: Velocity versus substrate concentration for the hydrolysis of PNPA by PZD2

Hofstee analysis of the data gives a K_m of 170 ± 20 μM, k_{cat} of 4.6 ± 0.2 × 10^{-4} sec^{-1}, and k_{cat}/k_{uncat} of 180 where k_{uncat} (2.5 × 10^{-6} sec^{-1}) is the rate of hydrolysis under the same conditions. The kinetics of PZD2 are comparable to those of the first catalytic antibodies: K_m of 208 μM and. k_{cat} /k_{uncat} of 770 for MOPC167, and k_m of 1.9 μM and k_{cat}/k_{uncat} of 960 for 6D4.

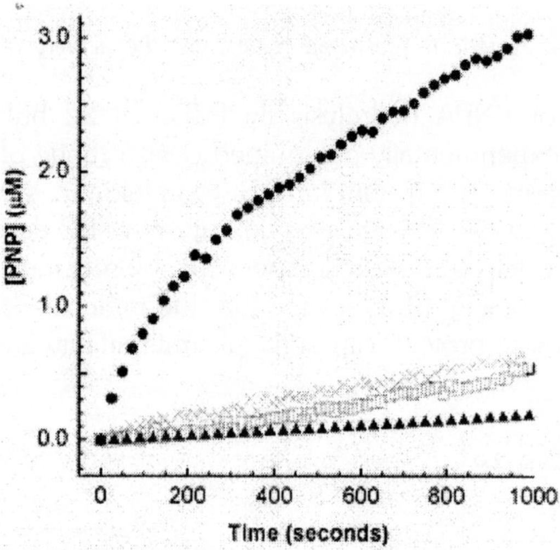

Figure 6: Buffer corrected hydrolysis of PNPA by PZD2 (•), PZD2 H17A () wild type thioredoxin (▲).

De novo designed peptides have recently been designed which catalyze pnitrophenyl acetate hydrolysis by a second order mechanism (i.e., without substrate saturation). The second order rates for these peptides were determined under different conditions than those used here making direct comparisons difficult. At pH 5.1, rates of 0.29 Msecand 0. 056 Msechave been reported. Extrapolation based on the pH rate profilesreported for the hydrolysis of similar activated esters suggests rates of approximately 0.7 and 0.2

Msecfor p-nitrophenyl acetate hydrolysis at pH 7, respectively. The second order rate of hydrolysis by 4-methyl imidazole (determined under the conditions used in the analysis of PZD2) is 0.11 $M^{-1} sec^{-1}$. Computing a second order rate of hydrolysis for PZD2 as k_{cat}/K_m gives 2.7 $M^{-1} sec^{-1}$, which is a factor of 25 better than catalysis by 4-methyl imidazole and about an order of magnitude better than the peptide systems.

It is extremely important to note, however, that the lack of substrate saturation in the second order systems complicates this analysis since substrate affinity was determined for PZD2. Substrate affinity is central to enzyme catalysis and clearly will be important in the design of protein-based catalysts for more complicated reactions.

At high substrate concentrations, kinetics with an initial "burst" phase are a common feature of natural enzymes and are a consequence of a kinetic bottleneck on the reaction pathway. For nucleophilic hydrolysis of PNPA, the rate of PNP formation should decrease and plateau as the acyl-enzyme intermediate concentration approaches steady state. PZD2 displays burst phase kinetics at high substrate concentration (Fig. 6) consistent with the formation of an enzyme intermediate.

The observed burst phase kinetics could also be a result of slow product release; however, this is unlikely given the kinetic parameters of PZD2. Based on the burst phase data, he estimate $K_s = 1$ mM, $k_2 = 3 \times 10^{-3}$ sec^{-1}, and $k_3 = 8 \times 10^{-4}$ sec^{-1}. Analysis of the magnitude of the burst indicates that approximately 30% of PZD2 is in an active state.

Using the burst magnitude as a metric of the enzyme concentration indicates a k_{cat} of $1.5 \pm 0.1 \times 10^{-3}$ sec^{-1}, approximately three times larger than the kinetic analysis using enzyme concentrations determined from protein UV absorbance.

Wild type thioredoxin is essentially inactive, but does show weak second order PNPA hydrolysis consistent with its single surface exposed histidine at position 6. Mutation of the designed catalytic histidine to alanine in PZD2 (H17A) results in a protein with catalytic activity similar to wild type thioredoxin indicating that the designed catalytic histidine at position 17 is necessary for the enzyme-like activity in PZD2 (Fig. 6).

Mutation of the two other active site residues in PZD2 back to their wild type identities (A12F and A70Y) also results in a protein with activity similar to wild type. Additionally, at pH 5.7 the activity of PZD2 is almost entirely eliminated, consistent with protonation of the catalytic histidine (data not shown). The kinetic and mutational evidence strongly indicate that PZD2 is working as designed with H17 acting as a catalytic nucleophile and the space creating mutations (F12A and Y70A) forming a binding site for substrate.

According to its design, reaction of PZD2 with PNPA should yield an acylhistidine enzyme intermediate, increasing the population of a +42 mass unit species compared to free protein. Mass spectrometry of PZD2 clearly indicates increased population of an acylated species upon reaction with PNPA (Fig. 7A&B). Acylation due to reaction with PNPA is essentially absent for the H17A mutant of PZD2 (Fig. 7C&D), indicating that acylation is dependent on H17. From the data it is not possible to distinguish between direct acylation of H17, or H17 mediated (acid/base or acyl transfer) acylation of a different residue.

Naturally occurring enzymes are frequently inhibited in a competitive manner by inert compounds resembling the substrate molecule. With this in mind, he assayed the inhibitory effects of p-nitrophenyl glycerol (PNPG), a highly soluble compound with structural similarity to PNPA. A double reciprocal analysis of PZD2 catalyzed PNPA hydrolysis in the presence and absence of inhibitor shows the hallmark features of competitive inhibition (Fig. 8) implying that PNPG is able to bind in the active site and block PNPA binding.

A Dixon analysis, varying PNPG concentration at a fixed substrate concentration of 200 μM yielded a K_i of 20 mM. The decreased binding affinity of PNPG relative to PNPA may

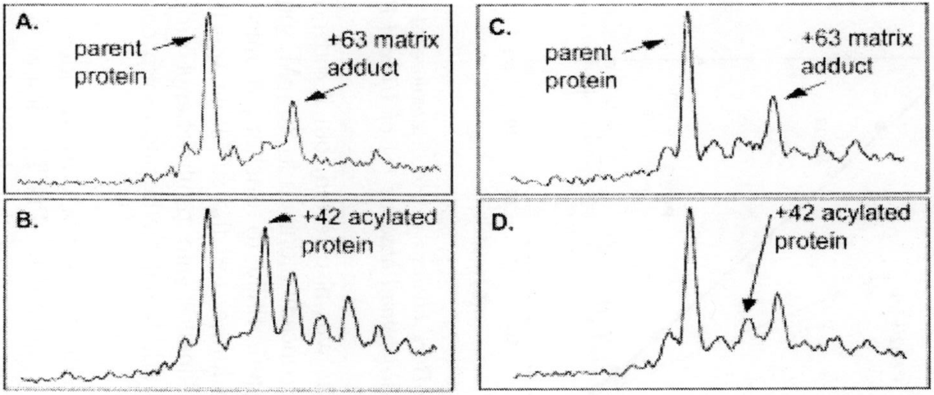

Figure 7: Trapping of an acylated intermediate by mass spectrometry: A) PZD2, B) PZD2 reacted with substrate, C) PZD2 H17A, D) PZD2 H17A reacted with substrate.

be due to burial of polar hydroxyl groups against hydrophobic regions in the active site or differences in the steric requirements for PNPG and PNPA.

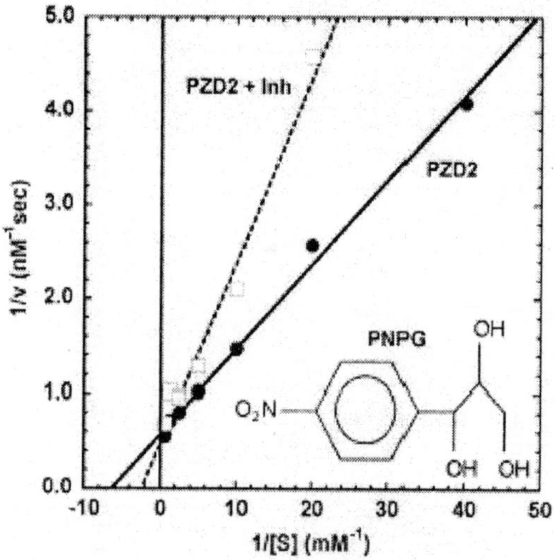

Figure 8: Lineweaver-Burk analysis of PZD2 catalyzed PNPA hydrolysis in the presence (□)and absence (•) of 10 mM PNPG.

p-nitrophenyl phosphate was also tested, but it did not show detectable inhibition at concentrations up to 20 mM, suggesting that PZD2 binds preferentially to uncharged nitrophenyl molecules. In order to further test the substrate specificity of PZD2 he assayed its kinetics with p-nitrophenyl propionate (PNPP), which has an additional methylene group compared to PNPA.

Hofstee analysis indicates within error an identical k_{cat}/k_{uncat} and a K_m of 110 ± 20 μM for PNPP, compared to 170 ± 20 μM for PNPA. Inspection of the designed structure of PZD2 with bound PNPA shows that space exists to accommodate the additional methylene of PNPP. Both favorable van der Waals interactions and hydrophobic burial afforded by the relatively open active site likely explain the slightly increased affinity of PZD2 for PNPP compared to PNPA.

Chapter 10

Role of Enzymes in Starch Hydrolysis

The digestion of starch is a complex mechanism. To optimise starch digestion it is important to take full account of the nature and form of the starch that is consumed. Commercially, starch is extracted in pure form from a variety of sources. Maize is the predominant source, but wheat, rice, potato, tapioca (manioc) and sago make a significant contribution. Other starch sources include barley, oats, yam, arrowroot and pea. Animal feed contains starch within plant tissues (especially cereal grains) while human foods are often supplemented with starch to provide desirable textural characteristics. The size, shape and granule size distribution of starch reflects the botanical origin as presented in Table 1.

Pure starch consists predominantly of α-glucan (ca. 99% of the dry matter) in the form of amylose and amylopectin. Amylose is a roughly linear molecule containing ~99% α-(1-4) and ~1% α-(1-6) bonds with a molecular weight of ~1x105-1x106 (Figure 1). Amylopectin (molecular weight ~1x107-1x109) is a much larger molecule than amylose and is heavily branched with ~95% α-(1-4) and ~5% α-(1-6) (Figure 1).

Each amylose chain is of the order of 1000 glucose units long whereas the unit chains of amylopectin range from ~12 to 120 anhydroglucose units (Figure 1). Starches are defined

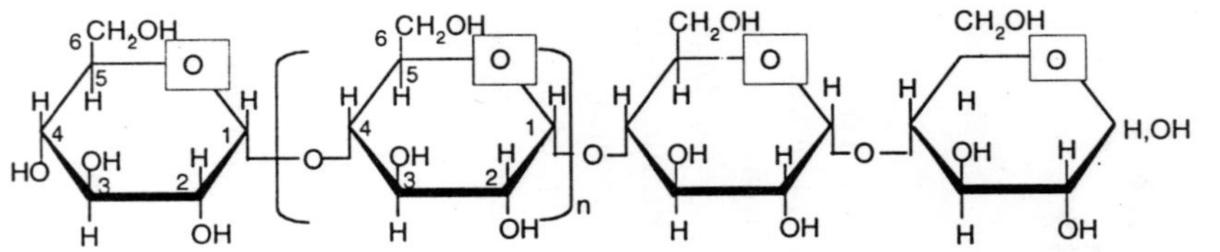

Amylose: α-(1.4)-glucan; average n = ca. 1000. The linear molecule may carry a few occasional moderately long chains linked α-(1→ 6).

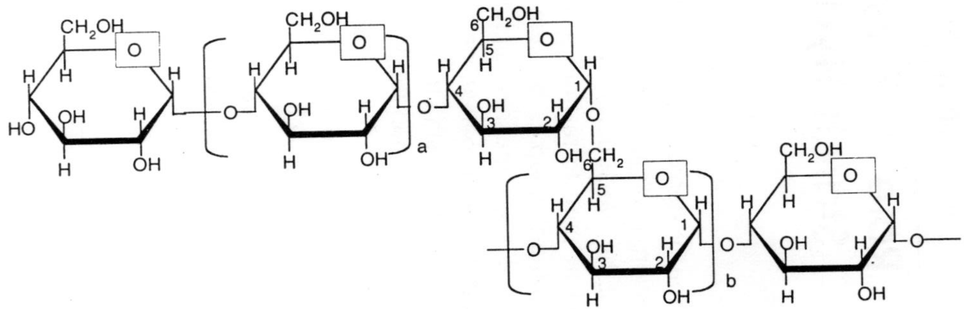

Amylopectin: α-(1→ 6) branching points. For exterior chains a = ca. 12-23. For interior chains = ca. 20 -120. Both a and b can vary according the botanical origin.

Figure 1: Basic structure of amylose and amylopectin

as waxy when the ratio of amylose to amylopectin is low (~15%), normal when amylose represents ~16-35% and high-amylose (or amylo-) when amylose exceeds ~36%. Waxy mutants are now commonly available for barley, maize and rice and have also been developed recently for wheat and potato. High-amylose mutants, predominantly from maize, are now available commercially.

Table 1: Characteristics of native starch granules from common sources.

Starch	Size (μm)	Shape	Distribution
Barley	2-5 (B-granules)	Spherical	Bimodal
	15-25 (A-granules)	Lenticular	
Maize (waxy and normal)	2-30	Spherical/ polyhedral	Unimodal
Amylomaize	2-30	Irregular	Unimodal
Millet	4-12	Polyhedral	Unimodal
Oat	3-10 (single)	Polyhedral	Unimodal
	80 (compound)		
Pea	5-10	Rentiform	Unimodal
Potato	5-100	Lenticular	Unimodal
Rice	3-8 (single)	Polyhedral	Unimodal
	150 (compound)		
Rye	5-10 (B-granules)	Spherical	Bimodal
	10-40 (A-granules)	Lenticular	
Sorghum	5-20	Spherical	Unimodal
Tapioca	5-35	Spherical/ Lenticular	Unimodal
Triticale	1-30	Spherical	Unimodal
Wheat	2-10 (B-granules)	Spherical	Bimodal

	15-35 (A-granules)	Lenticular

In commercially available purified starches moisture content ranges from 10 to 18% depending on the source, with cereal starches at the lower end and tuber starches at the upper end. Lipid is present in cereal starches only and is positively correlated with amylose content where it may represent ~1.5% of the granule. Lipid is probably all complexed with some of the amylose giving rise to 'lipid free' and 'lipid complexed' amylose.

The complexation of amylose with lipid affects the structure and functionality of starch granules (discussed below). Starch lipids (as distinct from surface lipid artefacts) comprise free fatty acids and lysophospholipids (LPL) in different proportions depending on the source of the starch. Purified starches contain low levels of protein (<0.5%) which largely represent the residues of biosynthetic enzymes involved in the synthesis of starch.

Starch granules are roughly spherical. These granules are synthesised by plants as semi-crystalline matrices, where crystallinity is generated by registration of amylopectin double helices into crystalline lamellae interspersed with amorphous lamellae comprising α-(1-6) branch regions of amylopectin and amylose (Figure 2). In high-amylose starches amylose probably forms double helices and crystalline domains accordingly.

In Figure 3 it is apparent that the alternating crystalline-amorphous lamellae provide the basis of semi-crystalline ('dark') growth rings which comprise about 16 crystalline lamellae, which are of approximately the same width as the interspersed amorphous ('light') growth rings (~140nm). Crystallinity is determined by wide angle X-ray diffraction scattering and is referred to as 'A-type' (found in most cereal starches), as 'B-type' (found in most tuber starches), and as C-type' (a mixture of both forms found for example in legume starches).

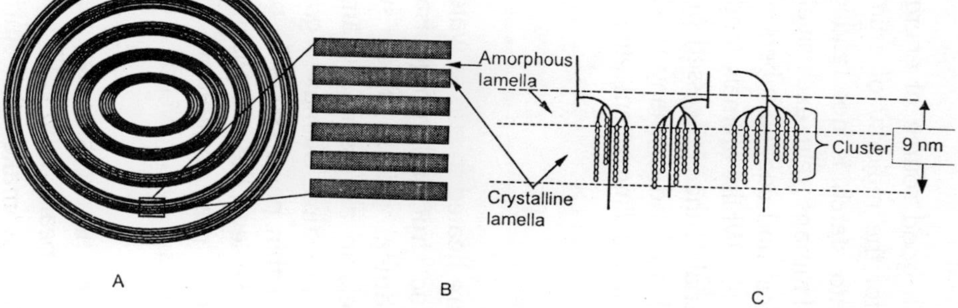

Figure 2: Diagrammatic representation of the lamellar structure of a starch granule. (A). Stacks of microcrystalline lamellae separated by amorphous growth rings. (B) Magnified view of the amorphous and crystalline regions. (C) Double helical structures formed by adjacent chains of amylopectin give rise to crystalline lamellae.

Plants synthesise starch in a compact semi-crystalline form to minimise storage space and maximise the concentration of energy stored. The underlying biochemical process is complex and incorporates a number of integrated enzyme systems. Clearly starch provides energy for growth and development when required. Animals consume starch in its native (raw) form.

Cereal grains are a good source of energy (referred to as 'starch equivalent') for the majority of animals, including poultry. However, potato starch is not readily digested by monogastric animals. Humans rely exclusively on cooked cereals (bread, pasta, rice) and cooked tubers and roots, as a source of digestible starch. Whilst animals tend to eat starch as part of a plant material (either processed or unprocessed) humans also add starch to food in many modified forms to provide texture (thickener, gelling agent etc.) and inevitably energy (calories).

Gelatinisation

When native starches are heated in excess water the granules undergo a characteristic structural reorganisation which is specific to their botanical origin. They swell and are transformed progressively to an essentially amorphous form with loss of organised structure. Ultimately granule structure is completely lost and a thin paste (<~4%) or gel (>~4%) is formed. At the molecular level the event is initiated by water 'plasticising' amorphous rearrangements and ultimately hydrating double helices as they unravel as a consequence of the elevated temperature.

Typically this process (in excess water) begins at ~45°C (onset temperature or T_o), peaks at 60°C (T_p) and is complete at 75°C (T_c). This process makes the starch completely digestible by starch hydrolysing enzymes. The gelatinisation process is shown schematically in Figure 3. It should be noted that gelatinisation is restricted if water is limiting.

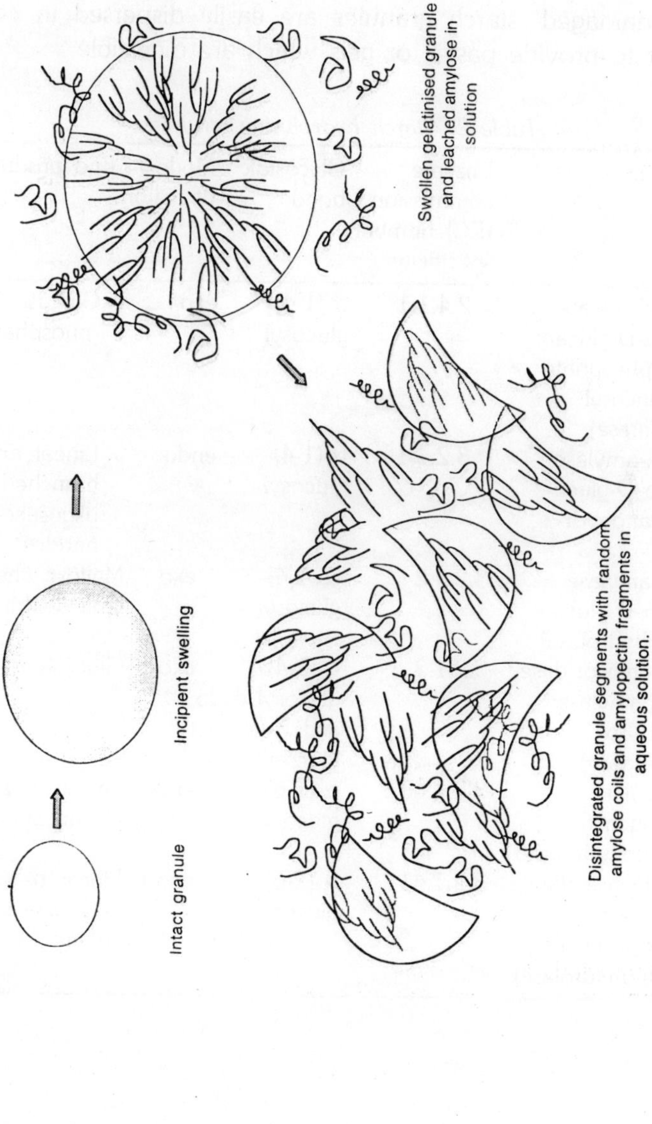

Figure 3: Idealised diagram of the swelling and gelatinisation of a starch granule in the presence of water

Modified Starches

Native starches are both indigestible and lack the capacity to contribute to structure in many foods. For this reason they are modified chemically, enzymatically or physically and thus acquire desirable properties for specific applications. Physical modification can be effected by attrition (e.g. milling) whereby the 'damaged' starch granules are easily dispersed in cold water to provide pastes or gels which are digestible.

Table 2: Starch hydrolysing enzymes

Enzyme	Enzyme commission (EC) number specificity	Glycosidic bond	Mode of action	End products
Phosphorylase (1,4-α-D-glucan: orthophosphate α-Dglucosyl transferase)	2.4.1.1	α-(1-4)-glucosyl	exo	Glucose 1-phosphate
Alpha-amylase (1,4-α-D-gluca-nglucanohydr-olase)	3.2.1.1	α-(1-4)-glucosyl	endo	Linear and branched oligosacc-harides
Beta-amylase (1,4-α-D-glucan maltohydrolase)	3.2.1.2	α-(1-4)-glucosyl	exo	Maltose and limit dextrins
Amyloglucosidase (glucoamylase; exo-1,4-α-glucosidase)	3.2.1.3	α-(1-4)-glucosyl and α-(1-6)-glucosyl	exo/ endo	Glucose
Isoamylase (glycogen 6-glucanohydrolase)	3.2.1.68	α-(1-6)-glucosyl	endo	Linear α-(1-4) glucan chains
Pullulanase (limit dextrinase; amylopectin 6-glucanohydrolase)	3.2.1.41	α-(1-6)-glucosyl	endo	Linear α-(1-4) glucan chains

In commercial practice, however, 'pre-gelatinised' starch is obtained by the simultaneous gelatinisation and drying of aqueous starch dispersions (e.g. by roller drum-drying). This physically modified starch gels instantly on the addition of cold water. Enzymes are used to hydrolyse starches particularly for the production of dextrins and glucose. Although α-amylase and amyloglucosidase are most often used in this respect to make glucose syrup, maltodextrins and crystalline glucose (dextrose), specialist dextrins and maltose may be obtained with β-amylase.

Enzymes responsible for starch hydrolysis are presented in Table 2. Chemical modifications are used to enhance desirable characteristics of starches. These include 'cross-linking' to provide chemical and thermal resistance and the addition of specific chemical moieties to prevent recrystallisation (retrogradation) of starch gels.

Structure of Digestive System

There is variation in the structure of the digestive system of poultry (birds) and mammals although the following description provides an overview of the starch digestion process which is relevant to mammals (especially man) and poultry. Specific differences in the respective digestive systems are not highlighted.

Animals and humans produce a range of digestive enzymes which hydrolyse starch. The saliva of both humans and poultry contains α-amylase, which hydrolyses accessible starch, but since this starch remains in the mouth for such a short period, the saliva causes little modification. The pancreas plays a major role in starch digestion where accessible starch in the duodenum receives pancreatic α-amylase into the lumen to hydrolyse α-(1-4) bonds and thus produce glucose sugars, oligosaccharides and dextrins.

Although glucose can be absorbed from the small intestine (especially terminal end of the duodenum and jejunum), maltose and dextrins generated from starch hydrolysis cannot. Absorptive epithelial cells which line the intestinal villi produce a number of other 'brush border' enzymes to allow for digestion and subsequent absorption of such materials (and disaccharides derived from other sources).

These enzymes are not released into the lumen of the small intestine but are bound to the membrane of microvilli. Consequently, some carbohydrate hydrolysis occurs at the surface of epithelial cells in the small intestine rather than the lumen of the gut. Brush border enzymes include sucrase, converting sucrose to glucose and fructose, and lactase converting lactose to glucose and galactose, but these are not relevant to starch metabolism.

A trehalase (converts trehalose to two glucose molecules) is also present. However, maltase (converting maltose generated from a-amylase activity to glucose) and isomaltase (also called α-dextrinase) which hydrolyses α-(1-6) bonds of isomaltose and α-dextrins) continue the digestive process and thus convert available starch to glucose. A representation of amylase and debranching enzyme hydrolysis of starch is shown schematically in Figure 4.

Native granular starch is hydrolysed very slowly by α-amylases and amyloglucosidase. The substrate (starch molecules compacted inside the granules) simply cannot be readily accessed by digestive enzymes and consequently hydrolysis is delayed. In the human digestive system, native starch is one of the four types of 'resistant starch' which may be broadly defined as follows:

— Type I Resistant Starch where starch granules embedded in plant tissue resist digestion,

— Type II Resistant Starch where certain native starch granules like potato and banana are particularly resistant to digestion compared with other starches (e.g. wheat, barley),

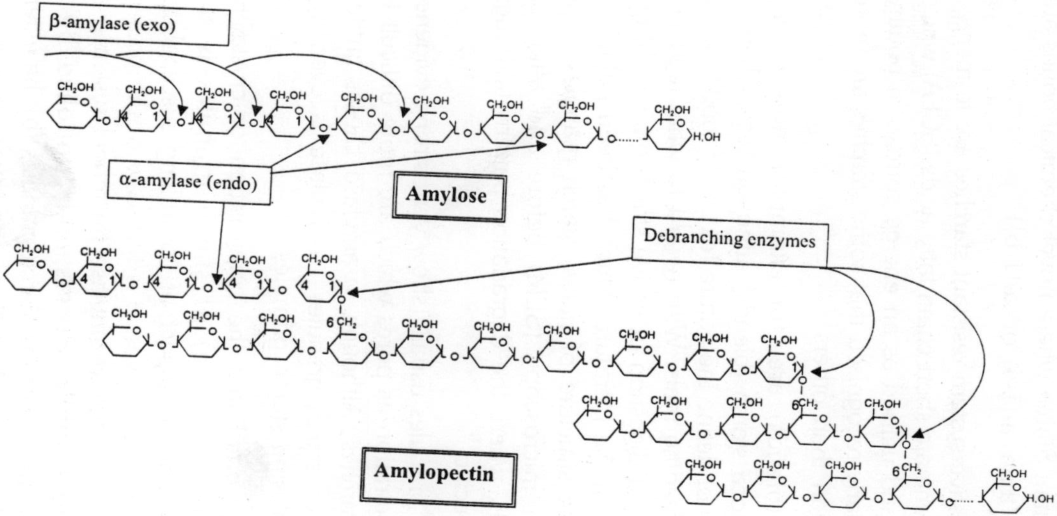

Figure 4: Action pattern of hydrolytic enzymes on amylose and amylopectin. β-amylase is an exo-acting enzyme cleaving successive β-maltose molecules from the non-reducing end of amylose or from the outer branches of amylopectin; it cannot by-pass α-(1-6) bonds. α-amylse is an endo-acting enzyme hydrolysing α-(1-4) bonds at random giving rise to malto-oligosaccharides (linear or branched, typically DP ~2-6); it does not hydrolyse α-(1-6) bonds. Debranching enzymes (e.g. isoamylase or pullulanase) specifically hydrolyse α-(1-6) bonds at the branching points of amylopectin.

— Type III Resistant Starch where starch molecules have recrystallised through hydrogen bonding (as in the staling of bread),

— Type IV Resistant Starch where novel chemical bonds are formed (other than a-(1-4) or a-(1-6)).

In the human digestive system resistant starches are fermented in the colon and generate short chain fatty acids (SCFA) which can be absorbed and utilised as an energy source. In poultry nutrition, it is similarly recognised that some starches are more resistant to digestion than others.

Pea starch (blends) are, for example, more slowly digested than tapioca starch (blends) although researchers in this field often do not define the nature (extent of modification) employed in the feeding trials. With regard to what regulates starch hydrolysis and digestion of native starch granules there is often confusion in the literature. Much has been published with respect to the structure of native starch granules using scanning electron microscopy (SEM) before and after α-amylase hydrolysis, where the degradation pattern (although slow) is starch specific.

Maize starch granules usually show very strong evidence of α-amylase penetration as pores which are more difficult to detect in other starches. Although α-amylase causes limited hydrolysis of native starch granules, β-amylase is even less effective in hydrolysing starch granules.

It is often reported that the composition or crystalline structure (A-, B- or C-type) of starch granules is the predominant regulator in respect of controlling susceptibility to hydrolytic enzymes. However, it is the view of the present authors that hydrolysis with α-amylase (in model systems) at least is a far more subtle process and controlled by a combination of (a) the size of the starch granules (b) the extent of damage (which is equivalent to gelatinisation in terms of generating amorphous material which is relatively easily hydrolysed by amylases and amyloglucosidase) as discussed elsewhere and (c) composition and structure.

Clearly if granules are large they have a smaller surface area to volume ratio and one would predict that amylase binding and potential hydrolysis would be less than for smaller granules as described above. Regarding granule composition, waxy starches are more crystalline than normal starches and are more readily damaged by milling, making them consequently more easily hydrolysed by α-amylase after the same amount of mechanical damage by milling. However, in native granules there is little difference in susceptibility to enzymes over the short term as the internal structures largely resist enzyme hydrolysis and, since the granules are approximately the same size, there is little capacity for hydrolysis.

For starch to be gelatinised and provide an accessible substrate for α-amylase hydrolysis the gelatinisation process must not be restricted. If water is limiting the amount of gelatinised starch is restricted together with the capacity of α-amylase to hydrolyse the starch. Similarly, the presence of non-starch polysaccharides during gelatinisation restricts the gelatinisation process and subsequently α-amylase hydrolysis.

In the digestive tract non-starch polysaccharides as cell wall material prevent starch digestion by preventing the digestive enzymes reaching the starch. This has been described elsewhere in the context of poultry nutrition although there is debate concerning the exact role of non-starch polysaccharides in this respect. It is also well known that feeds and foods contain anti-nutritional factors which again may be relevant in poultry nutrition.

Optimisation

For starch to be readily digestible it must be:

a. Amorphous (especially physically damaged or gelatinised) not crystalline

b. Freely accessible to digestive enzymes (not entrapped in food/feed particles),

c. In small particles or preferably solubilised,

d. Not associated with other molecules to form associations or complexes (e.g. amyloselipid complexes).

e. Not chemically modified in a form that prevents it from acting as a substrate for amylases.

When these criteria are met, the healthy digestive system has a greater opportunity to convert the starch fully to glucose.

Chapter 11

β-lactum Acylases in Protein Engineering

Recent years have brought new enzymes, new applications and new insights into the biochemical features of the β-lactam acylase family. The introduction of these enzymes as biocatalysts in industrial processes has resulted in new innovative processes for the production of fine chemi-als in the recent years. Driven by both environmental and economic considerations, particular progress has been made in the area of semi-synthesis of penicillin and cephalosporin antibiotics. Not only the replacement of the acyl side chain but also modifications of the other substituents of the β-lactam nucleus are being performed with specific enzymes today.

The enzymes of the β-lactam acylase class are traditionally used for the deacylation of the β-lactam compounds penicillin G and glutaryl-7-aminocephalosporanic acid (glutaryl-7-ACA) to 6-aminocephalosporanic acid (6-APA) and 7-ACA. These β-lactam nuclei are the building blocks used in the production of semi-synthetic β-lactam antibiotics. The most well known β-lactam acylase is the penicillin G acylase (PGA) from Escherichia coli. This enzyme hydrolyses the amide bond between the primary amino group of the β-lactam nucleus and the carboxyl group of the phenylacetic acid side chain of penicillin G.

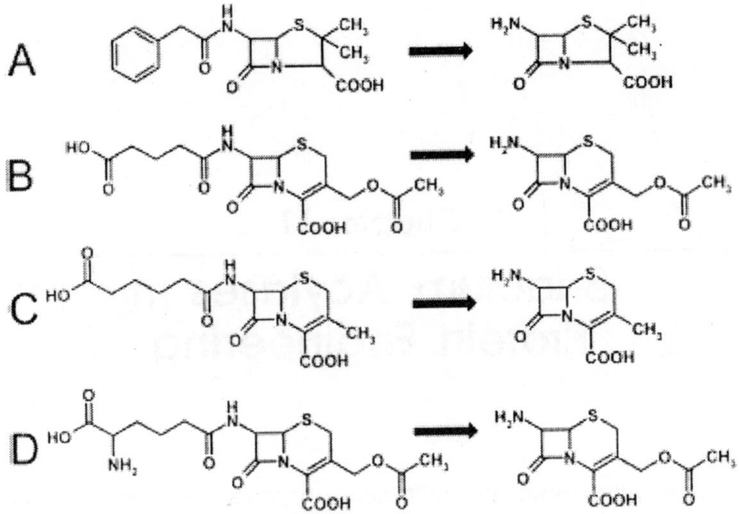

Figure 1: The hydrolysis of β-lactam fermentation products by â-lactam acylases

Likewise, cephalosporin acylases hydrolyze the amide bond between the nucleus and the aliphatic side chain of cephalosporins (Figure 1). The elucidation of the 3D structure of E. coli PGA has revealed a unique catalytic mechanism involving an N-terminal nucleophilic serine that is activated by the free N-terminal amino group. Although advances in the fermentation processes and the hydrolytic processes have been made, the most interesting manuscripts in the last few years have focused on protein engineering and directed evolution of β-lactam acylases.

Novel b-lactam Acylases

A newly characterized β-lactam acylase originates from Achromobacter sp. CCM 4824. The discovery of this enzyme was a result of classical screening of 2000 isolates from soil samples and the corresponding bacterium has been characterized on the basis of the 16S RNA sequence as being 99.6 % identical to Achromobacter xylosoxidans. The enzyme was found to hydrolyze β-lactams containing an a-amino group in the side chain, such as ampicillin and cephalexin, with a 2-fold higher activity compared to penicillin G. It is the only acylase described so far that is able to hydrolyze ampicillin faster than penicillin G, which is potentially interesting for the enzymatic synthesis of ampicillin, amoxicillin and cephalexin.

Furthermore, the substrate specificity of the acylase enzyme from Streptomyces lavendulae was recently investigated in detail. In accordance with its former classification this enzyme has a high activity on Penicillin V and almost no activity on Penicillin G. Surprisingly the highest activity was found towards substrates with octanoyl as the acyl side chain. These results taken together point to a highly hydrophobic active site, and it was proposed to classify this enzyme as a Penicillin K acylase.

Interestingly, a new class of β-lactam synthases was discovered with the amino-acid ester hydrolases of Xanthomonas citri and Acetobacter turbidans, which show a preference for substrates with a-amino side chains and have biochemical features distinct from the other β-lactam acylases. These enzymes are especially suited for synthetic reactions as the β-lactam amide formed is not a substrate. The identification of this new class of β-lactam amidase was a further investigation of ampicillin acylases that were first described 30 years ago. The 3D structure of this enzyme has been determined recently and will be discussed separately.

Within the past 1.5 years the number of available bacterial genome sequences has grown from 71 to over 155. Database mining for novel β-acylases has therefore become feasible. A recent BLAST search using the paradigm of β-lactam acylases, E. coli PGA, as a template, revealed over 50 homologous genes in the various database libraries. However, the functional analysis of all these new putative genes is very time consuming, and it is therefore expected that the number of well characterized novel acylases will increase steeply in the coming years.

Protein Engineering

Mutating the existing β-lactam acylases can improve the physicochemical and catalytic characteristics. In a rational approach, non-conserved basic surface residues of Bacillus megaterium PGA were mutated into the small, hydrophobic residue alanine. This resulted in a two to three-fold increased half-life at 55°C or in 40% DMF. Alternatively, the charged residues of E. coli PGA were mutated to lysines, thereby improving the interaction with the immobilization agent glyoxyl-agarose. The inactivation of the immobilized enzyme was decreased four to eleven-fold depending on conditions.

With regard to the synthesis of β-lactam antibiotics, it was found that mutating several active site residues of E. coli PGA increased the fraction of the new acyl side chain that is coupled to the β-lactam nucleus (i.e. the synthesis to hydrolysis ratio Vs/Vh), thereby increasing the yield of ampicillin and cephalexin synthesis two to four-fold. In a different approach, the V_s/V_h ratio was improved by DNA family shuffling, using E. coli, Kluyvera citrophila and Providencia rettgeri PGA with the P. rettgeri PGA gene as backbone. Chimeras exhibited a 40% improved ratio compared to P. rettgeri PGA, but the ratio was not significantly improved compared with E. coli PGA.

To increase the activity of cephalosporin acylase on the β-lactam fermentation product cephalosporin C (CPC; Figure 1), Kabsch and co-workers made a model of a cephalosporin acylase with bound glutaryl-7-ACA and CPC. Several combinations of mutations were proposed to improve the binding of the desired substrate, but experimental results of these mutations are yet to be published. Based on a different model, Kim and co-workers selected active site residues that clashed with CPC. Step-wise random mutagenesis of these residues resulted in an 8-fold increase in hydrolysis, which is, however, still too low for industrial applications.

Wim J. Quax and co-workers used directed evolution to alter the substrate specificities of cephalosporin acylase towards the novel β-lactam fermentation product adipyl-7-ADCA (Figure 1). Random mutagenesis followed by a growth selection on adipylderivatives identified three hot-spot residues in the gene, of which one, Asn266 (which corresponds to position 68 of the β-subunit, Asnβ68), was not identified as an possible target for mutagenesis from the 3D structures. The catalytic efficiency of the selected mutants towards adipyl-7-ADCA was increased two to four-fold.

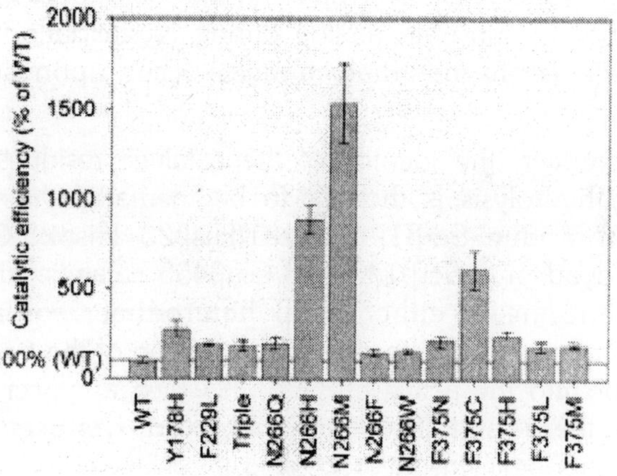

Figure 2: Mutagenesis alters the substrate specificity of a cephalosporin acylase

Subsequently, the effect of two of the three hot-spot residues on activity was thoroughly analyzed through the incorporation of all 20 natural amino acids at these positions, showing several mechanisms to improve activity (Figure 2). Substitutions of Asn266 with methionine afforded an 11-fold improved catalytic efficiency (Figure 2), and, moreover, improved activity towards CPC. In contrast to these results, growth selection of mutants of E. coli PGA on glutaryl-leucine resulted in variants with an improved interaction with the leucine moiety of the selection substrate, but no activity towards the corresponding β-lactam compound, glutaryl-7-ACA.

Crystal Structures

Recently, crystal structures of several highly homologous cephalosporin acylases have been resolved. Although the structures are all virtually identical, some controversies remain. The structure and model of glutaryl-7-ACA bound enzyme are very similar to the structure of the native enzyme, suggesting a lock-and-key mechanism for substrate binding. However, the binding of the substrate analog bromoacetyl-7-ACA to the non-solvent accessible residue Trpβ4 and the subsequent hydrolysis of the bound substrate analog during crystallization suggest that large conformational changes occur upon substrate binding.

Moreover, the identity of the catalytic residues is also unclear; hydrolysis is thought to be mediated via a single catalytic residue Serβ1, a Serβ1/Hisβ23 Hisβ23/Gluβ455 double dyad or a Serβ1/Hisβ23/Gluβ455 catalytic triad. The active enzyme is either a aβ heterodimer, or a (aβ)2 heterotetramer. The autocatalytic processing of the transcribed precursor into the active enzyme, in which a spacer peptide between the a- and β-subunit is removed, has been studied via mutants deficient in processing.

It was shown that a bound water molecule plays a pivotal role in activating the hydroxyl group of Serβ1, which is the nucleophile in the first processing step (Figure 3). If this water is forced out by mutating Pheβ177 to proline, the hydroxyl group is not positioned correctly and processing is inhibited. In the nondeficient enzyme, the formed intermediate is thought to be hydrolyzed by either the same water molecule or via a different one following a structural rearrangement of the spacer peptide.

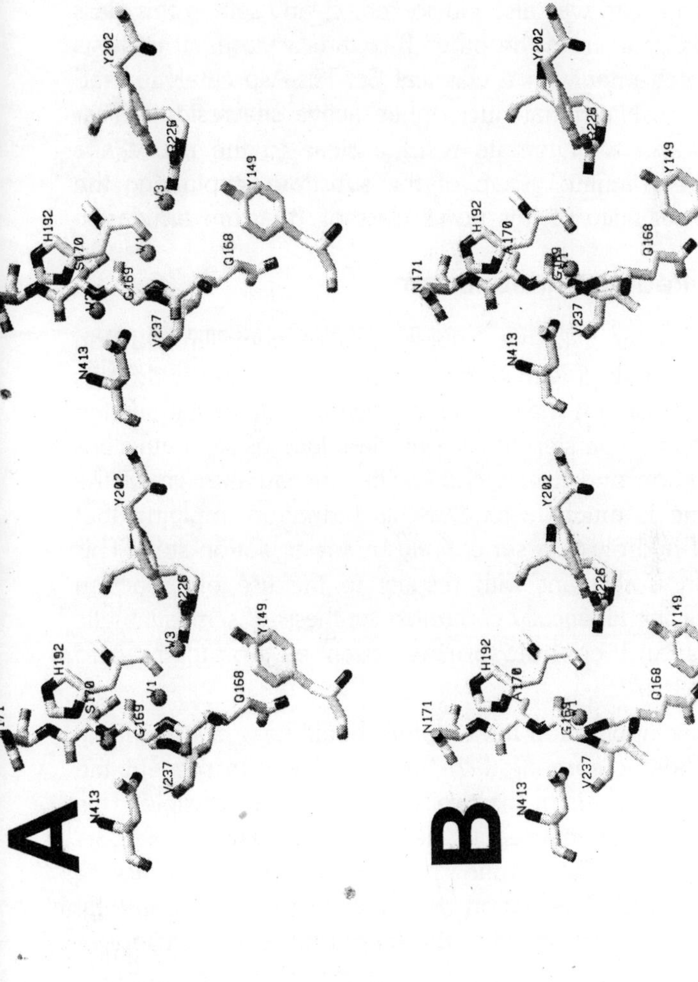

Figure 3: Two proposed mechanisms for the autocatalytic processing of cephalosporin acylase

Finally, different lengths of the removed spacer peptide have been reported for the different homologs of cephalosporin acylase, ranging from eight to eleven residues. This clearly indicates that more work is yet to be done. The crystal structure of the a-amino acid ester hydrolase from Xanthomonas citri was also elucidated, clearly setting this class of enzymes apart from the other β-lactam acylases. It consists of a homotetramer with a classical Ser-His-Asp catalytic triad instead of an Nterminal nucleophile active site residue. Two aspartates and a glutamate residue close to the active site stabilize the a-amino group of the substrate, explaining the substrate specificity of this novel class of β-lactam amidases.

Acylase Reaction Mechanism

In the past year a kinetic analysis of the penicillin reaction steps was published with emphasis on not only the hydrolysis reaction but also on the synthetic reaction. The acylation step and the deacylation step have been described as separate steps in the reaction and it was shown that the breakdown of the acyl-enzyme is much faster than its formation implying that the rate of hydrolysis is set entirely by the acylation step. This information is relevant with respect to the use of β-lactam acylases for the kinetically controlled synthesis of semi-synthetic penicillins and cephalosporins, such as ampicillin and cephalexin.

For this application the enzyme should have a low affinity for the antibiotic compared to the acyl donor to prevent the competing hydrolysis of the product as much as possible. This knowledge has been used to design novel variants of PGA with an improved V_s/V_h ratio. It appears that the kinetics of the acylation and deacylation step can be changed separately, but it should be stressed that the thermodynamic equilibrium can not be changed by protein engineering of the enzyme.

The conditions for optimal yield of the antibiotic product can be influenced by varying conditions (pH) and the concentrations of the substrates. The complexity of these processes has been analyzed elegantly by a neural network approach. To circumvent kinetic limitations, efforts have also been directed at precipitating the antibiotic product to drive the reaction towards one end of the equilibrium. A theoretical description of this heterogeneous "aqueous solution-precipitate" system was found to match quite accurately the experimental conversion of 86 % and 76 % for ampicillin and amoxicillin, respectively.

In a further refinement, the feasibility of precipitation-driven synthesis of a zwitterionic β-lactam antibiotic (e.g. amoxicillin) has been investigated. For amoxicillin it was found that Zn^{2+} ions drive the precipitation of the product. In experiments where $ZnSO_4$ was added, much higher levels of amoxicillin were reached, but an undesirable side-effect of increasing concentrations of Zn^{2+} ions was that β-lactam degradation seemed to increase.

Industrial Applications

Although the enzymatic production of amide bonds can be carried out with unmodified substrates following a thermodynamically controlled strategy, the situation is very different using aliphatic amines or β-lactam nuclei. Whereas the pK of the amino group of the β-lactam nuclei is below pH 5, the pK of aliphatic amines is around 9.5 or even higher. This high pK value of the amine makes it necessary to either use a high pH value to obtain a high percentage of nonionized amine, or to use a high concentration of organic solvent to shift the pK of the carboxylic acid. This requirement has reduced the number of possibilities for performing a wide range of amidation reactions, including antibiotic synthesis.

Recently, some significant improvements in the design of thermodynamically controlled syntheses catalyzed by PGA have been reported. On the one hand, the screening of different co-solvents showed some of them to have a similar effect on the pK of the carboxylic acid, but a low deleterious effect on the PGA stability. A reaction in acetonitrile containing

8% water afforded ampicillin in up to 86% yield using the more stable PGA from Alcaligenes faecalis.

Also, the yield of cephalexin could be increased significantly by using an aqueous two-phase system containing 20% (w/w) polyethyleen glycol (PEG) 400 and 15% (w/w) magnesium sulfate, because the inhibiting product accumulated separate from the enzyme in the PEG phase. On the other hand, immobilization of the enzyme on solid supports increases the robustness of the enzyme towards organic media, non-physiological pH and temperature. High yields of cephalexin were obtained using immobilized PGA, but the process needs further refinements.

The latest addition to the list of applications of the β-lactam acylases is enantioselective acylation/hydrolysis. Whereas the side chain binding site of β-lactam acylases is restrictive, the aminic subsite accepts a large number of structures, allowing for the resolution of a large number of compounds. Analysis of the crystal structure of E. coli PGA identified hydrophilic and hydrophobic parts in the substrate-binding site, restricting the binding of aromatic substrates.

Models of transition-state analogs of aromatic amino acid methyl esters showed that the ester group of the L-enantiomer can form an additional H-bond that cannot be formed by the ester of the D-enantiomer, explaining the observed enantioselectivity. By contrast, binding of aliphatic amino acids allowed for more degrees of freedom; esters of both enantiomers could form the H-bond, explaining why no enantioselectivity was observed.

E. coli PGA specifically acylates the R-enantiomer of phenylglycinonitrile with phenyl acetic acid (PAA). A low solubility of the product makes the reaction virtually irreversible and stoichiometric, leaving a high enantiopure S-phenylglycine in solution. An extensive study of the enantioselective acylation of a β-amino acid ester showed that the source of the immobilized enzyme, pH, concentration and type of PAA and amine, and the reagents themselves (PAA or ethyl-PAA) influenced yield and enantioselectivity of ethyl 3-amino-5-(trimethylsilyl)-4-pentynoate to a great extent.

Highly pure free amine and amide was obtained using 50 g/L phenylacetic acid as free acid, 100 g/L ethyl-3-amino-5-(trimethylsilyl)-4-pentynoate as the amine and 14,000 U/L PGA-450 as enzyme at pH 5.7, 28°C. Moreover, the reaction could be scaled up to 70L batches, with a half-life of the enzyme of approximately 20 runs. In addition, enantioselective hydrolysis of an ester lacking a PAA moiety by E. coli PGA was reported, as well as enantioselective hydrolysis of the amide bond or ester bond of several glutaryl coupled substrates by a cephalosporin acylase. This greatly increases the substrate range for enantioselective reactions catalyzed by β-lactam acylases.

Other new directions of research include the incorporation of immobilized PGA in a polyacrylamide gel (allowing for the control of penicillin G hydrolysis by applying a specific current), the coupling of PGA to an HPLC column resulting in an HPLC enzyme reactor, and solid-state ampicillin synthesis by E. coli PGA in a wet powder of reagents, immobilized enzyme, salt hydrates and a little as 10% water resulting in high space-time yields.

Chapter 12

Applications of Cross-linked Epoxide Hydolase

Cross-linked enzyme crystals (CLECs) are formed by cross-linking enzyme microcrystals using a bifunctional reagent. CLECs have been shown to possess increased stability in the presence of organic cosolvents and at high temperatures compared to the respective enzymes in solution and have therefore become attractive as a form of immobilised biocatalyst for use under reaction conditions which may cause loss of biocatalyst activity. Compared to soluble enzymes, CLECs have the additional advantage of being easy to recover and recycle. CLECs also generally have a higher specific activity by weight than immobilised enzymes where the carrier makes up a large proportion of the catalyst.

In this chapter, Margolin and Navia have given a comprehensive review of CLEC technology applied to biocatalysis. Some other details as follows. CLECs of lipases, proteases and esterases are some of the enzymes currently sold for biocatalytic application.

Recently, the potential for using CLECs as a selective crystalline chromatographic material for chiral separation has been demonstrated. Cross-linked protein crystals can also be used for the controlled release of therapeutic proteins in medical applications, since crystal dissolution can be carefully controlled by the method of cross-linking and the nature of the incubation environment. The remarkable reported stability of CLECs prompted us to evaluate the possibility of forming cross-linked crystals of epoxide hydrolase for application in the kinetic resolution of racemic epoxides.

The enantioselective hydrolysis of racemic epoxides by an epoxide hydrolase from Agrobacterium radiobacter AD1 can be carried out at high substrate concentrations in a biphasic liquid system. It has been shown, however, that in such a system epoxide hydrolase inactivation can occur by direct contact of the enzyme with the organic liquid and by accumulation of the formed diol product in the aqueous phase at high concentrations.

Disadvantages recognised in the use of CLECs include the lower specific activity compared to free enzyme. This is sometimes attributed to reduced mass transfer rates of substrates and products in the crystal channels or to alteration in enzyme conformation during crosslinking. In immobilised enzyme systems, mass transfer limitations can reduce the observed enzyme selectivity.

The resolution of racemic styrene oxide by wild-type epoxide hydrolase is an example of this, where the preferred (S) substrate enantiomer inhibits the conversion of the ®-enantiomer which is converted last. Mass transfer limitations would result in a reduction in yield of the remaining epoxide substrate enantiomer, which is clearly undesirable. To reduce the impact of potential mass transfer limitations in this work, they used an active site mutant of this epoxide hydrolase (Tyr215Phe) which converts several substituted styrene epoxides with a much higher intrinsic enantioselectivity than the wild-type enzyme.

For some substrates e.g. p-chlorostyrene oxide, the rate of conversion of the second enantiomer is reduced virtually to zero so that mass transfer limitations would have little effect on effective enantioselectivity of the reaction. The reduced specific activity observed with the mutant enzyme would also tend to make the system kinetically limited rather than mass transfer limited, even at the high enzyme concentrations encountered in CLECs. In addition, it has been shown that the Y215F epoxide hydrolase mutant is resistant to inactivation by vicinal diols at high concentrations which makes it a preferable for use in large-scale applications.

Batch crystallisation is the most suitable method to obtain large amounts of protein crystals. Screening for the best conditions for crystallisation is carried out in the same manner as in the production of crystals for X-ray crystallographic studies. Conditions such as temperature, pH, precipitant type and concentration, protein concentration and purity and salt concentration are tested using the hanging drop crystallisation method. These conditions are then used as first guesses for batch crystallisation trials, and can be further optimised by experimental design. In batch crystallisation, seeding and rate of mixing are additional parameters that affect the crystallisation process.

Crystallisation should be optimised to maximise crystallization yield and most importantly the crystallisation should be reproducible and amenable to scale-up. Crystal properties which impact the functionality of the CLECs are size and uniformity of size. In general, large crystals of thickness greater than 5 m can potentially suffer from mass transfer limitations. The length of the crystals on the other hand has to be large enough (≥150 μm) to ensure good recovery of CLECs by filtration to avoid large losses of CLEC material during recycling.

Cross-linking of enzyme crystals is often carried out using the homobifunctional reagent glutaraldehyde which reacts with the ε-amino groups of lysine. The inter- and intramolecular cross-linking helps to resist unfolding at high temperatures and in unfavourable environments. Glutaraldehyde can form oligomers of different length and therefore can form intramolecular cross-links spanning existing distances between lysine residues, reducing possible distortion of the protein conformation.

The reduction in specific activity of CLECs, which is frequently observed compared to enzyme in solution, is often explained by a decrease in conformational flexibility of the enzyme in the crystalline state. In addition, the catalytic mechanism may be altered in the formation of CLECs if an essential residue is modified during crosslinking. The degree of cross-linking is determined by the cross-linking reagent concentration, cross-linking time, temperature and pH. A trade-off between stability of the crystals and activity has been reported as a function of the degree of cross-linking and the optimal degree of cross-linking seems to depend on the particular protein and crystal form.

Chemicals for Enzyme Preparation

The Y215F epoxide hydrolase was produced using a recombinant *E. coli* strain and purified by a two-step chromatography procedure as described by Rink et al. An initial purification step was carried out using a DE52 anion exchange column. The second purification step involved the use of a hydroxylapatite column. The purified enzyme was dialysed against TEMA (25 mM Tris-SO_4, 1 mM EDTA, 1 mM mercaptoethanol, 0.02% sodium azide) buffer at pH 7.5.

Racemic styrene oxide (98%), (*R*)-styrene oxide (93% e.e., determined by GC) and (*S*)-styrene oxide (99% e.e., determined by GC) were obtained from Sigma-Aldrich. *p*-Nitrostyrene oxide (pNSO) was synthesised as described elsewhere, and (*R*)- (99% *e.e.*, determined by HPLC) and (S)-pNSO (99% *e.e.*, determined by HPLC) were obtained by separation of the enantiomers by preparative chiral HPLC. Glutaraldehyde, PEG 8000, NaCl, and imidazole were obtained from Sigma.

Analytical Procedures

Samples for the determination of styrene oxide concentrations were prepared by withdrawing 50 l of the reaction mixture and extracting with 300 μl of diethylether containing mesitylene as internal standard. Chiral analysis was carried out with a Helwett-Packard 5890 gas chromatograph with an FID-detector, using a Chirasil Dex CB (Chrompack) column. Samples for the determination of pNSO concentrations were prepared by withdrawing 50 μl of the reaction mixture and extracting with 250 μl of noctane

containing mesitylene as internal standard. The samples were analysed by chiral HPLC on a Chiralpak AS (Daicel) column with UV detection at 254 nm. Isocratic elution with hexane:2-propanol (ratio 9:1) was carried out at a flow rate of 1.3 ml min^{-1} (retention time (*R*)= 10.5 min and retention time (S) = 15.7 min).

Batch Crystallisation

Enzyme was crystallised by using a batch crystallisation method. The crystallisation conditions which had been used successfully to crystallise the Y215F epoxide hydrolase mutant by the hanging drop method were used for the initial screening. Optimal crystallisation conditions were screened by varying the concentration of enzyme and polyethyleneglycol (PEG 8000) in the crystallisation mixture. Enzyme concentration was varied between 2 mg ml^{-1} and 5.3 mg ml^{-1} and the PEG 8000 concentration was varied between 16 and 26% w/v. Imidazole and NaCl concentrations were fixed at 100 mM and 200 mM, respectively, and the crystallisation was carried out at pH 8.4 and a temperature of 22°C. The screening was carried out using an Oryx 6 crystallisation robot using crystallisation liquor droplets of 2.2 l total volume.

At the 1 ml scale, crystallisation was carried out in solution containing 100 mM Imidazole (pH 8.4), 200 mM NaCl and 24% w/v PEG, with an enzyme concentration of 4 mg ml^{-1}. All the crystallisation liquor components were mixed thoroughly together before adding enzyme to the desired concentration. Crystals were grown at 22°C.

Cross-linking of Crystals

Cross-linking of crystals was carried out with glutaraldehyde. Crystals were transferred (using a micropipette) to 200 µl liquid of the same composition as the crystallisation liquor which also contained glutaraldehyde at a concentration of 5% v/v (25% stock solution). Crosslinking was carried out in wells of a 96-well plate which was UV grade allowing the visual inspection of the CLEhCs through a microscope using polarised light. The crystals were allowed to cross-link for 0.5 h, 1 h or 24 h. The CLEhCs were washed five times with 400 l TEM buffer (50 mM Tris-SO_4, 1 mM EDTA, 1 mM mercaptoethanol, pH 7.5) by repeatedly adding fresh buffer to the plate well, mixing by repeated injections with the Gilson pipette, allowing the crystals to settle and removing the wash liquid with a Gilson pipette. The CLEhCs were stored at room temperature until required.

Activity Assays

Activity assays were carried out by following the hydrolysis of the colorimetric substrate pNSO at 310 nm. The extinction coefficients for the epoxide and the corresponding diol are 310 = 4289 $M^{-1}cm^{-1}$ and 310 = 3304 $M^{-1}cm^{-1}$, respectively. Assays were carried out in 96-well plates to enable several reactions to be carried out simultaneously. The small well volume made it possible to measure reasonable reaction rates with very small amounts of CLEhCs. The plates were read using a PowerWaveX plate reader with periodic shaking during readings. In each well, the reaction mixture consisted of 200 µl TEM buffer, pH 7.5 and a similar amount of CLEhCs. Assays were started by adding of 1 l of a 30 mM stock solution of pNSO dissolved in acetonitrile. Activity was determined by calculating initial reaction rates for no more than 20% substrate conversion. CLEhC reusability was tested by repeating assays, performed as described above, using the same batch of CLEhCs in the reaction mixture.

Enantioselectivity

Kinetic resolutions were performed with racemic pNSO and racemic styrene oxide at initial substrate concentrations of 1 mM. Enzymatic conversion of the enantiomers of each epoxide was also tested separately at initial concentrations of 0.5 mM. For each epoxide a dedicated batch of CLEhCs was used for the kinetic resolution reaction followed by the single enantiomer conversions. The batches were washed thoroughly with TEM buffer between reactions. The reaction mixtures consisted of 1.5 ml TEM buffer (pH 7.5) containing a similar amount of CLEhCs.

The reactions were carried out at room temperature (22°C) and were started by adding pNSO from a stock solution in acetonitrile (final acetonitrile conc. < 1% v/v), or styrene oxide from a concentrated aqueous solution, to the desired initial substrate concentration. All incubation vials were stoppered and shaken using a mechanical vortex shaker at a fixed rotational speed which maintained suspension of the crystals in solution. Substrate conversion was followed by periodically taking 50 µl aliquots of the reaction mixture and the samples were analysed as described above. To avoid withdrawing CLEhCs while sampling, the mixing was stopped and crystals allowed to settle during 1 min before withdrawing liquid.

CLEhC stability

Stability of the CLEhC preparation was tested by performing activity assays in the 96-well plates using the colorimetric substrate pNSO as described above. Stability was determined by measuring the change in activity of a CLEhC batch before and after incubation under a given set of conditions. Activity was determined by calculating initial reaction rates for no more than 20% substrate conversion. In this way, CLEhC themostability was tested by incubating the CLEhCs at 60°C and measuring activity after 1.5 h and then 5 h. After the first activity assay the CLEhCs were washed with TEM buffer before re-incubation at 60°C. Similarly CLEhC stability was tested during incubation with 50% DMSO and during incubation in solution containing 100 mM 3-phenoxy-1,2-propanediol.

Soluble Epoxide Hydrolase Thermostability

Thermostability of soluble Y215F was measured by incubating a 0.2 mg ml^{-1} solution of the mutant in TEM buffer at 50°C or 60°C and following the change in enzyme activity with time. The thermostability of the wild-type epoxide hydrolase was similarly assessed at 50 and 60°C by incubation in TEM buffer at the same concentration as the mutant enzyme. Periodically, enzyme activity was assayed by following the conversion of the colorimetric substrate pNSO as described elsewhere. Samples from the enzyme incubation were diluted with TEM buffer so that the assay mixture reached room temperature before the addition of pNSO.

Product Inhibition

The inhibitory effect of diols on CLEhC activity was determined by measuring the change in CLEhC activity between incubation in TEM buffer and incubation in diol solutions at different concentrations. The activity of a CLEhC batch was determined by the colorimetric assay as described above. After the assay, the CLEhC batch was washed with TEM buffer 5 times and CLEhCs were resuspended in a diol solution, whereupon the activity assay was repeated in the same manner. The change in activity was expressed as a percentage change from the activity in TEM buffer.

Results

Batch Crystallisation

The remarkable stability commonly reported for CLECs prompted us to evaluate the use of cross-linked crystals of

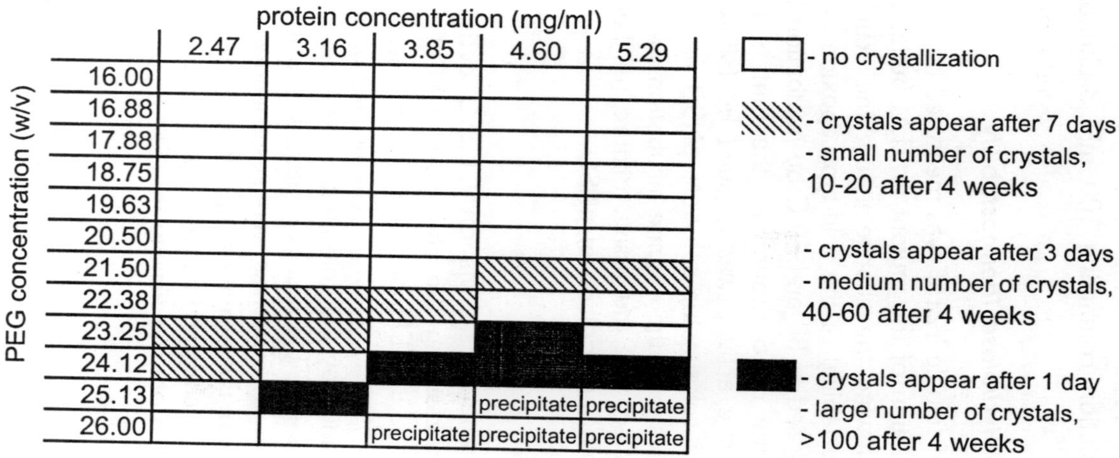

Figure 1: Screening of batch crystallisation conditions for Tyr215Phe mutant epoxide hydrolase. PEG 8000 and protein concentrations were varied as shown, with fixed imidazole (100 mM; pH 8) and NaCl (200 mM) concentrations. Screening was carried out using an Oryx 6 crystallisation robot with crystallisation liquor droplets of 2.2 μl total volume

epoxide hydrolase for the kinetic resolution of racemic epoxides. Batch crystallisation is a crystallisation method which is generally amenable to scale up and was therefore used to obtain crystals of the epoxide hydrolase. Batch crystallisation conditions were screened by varying enzyme and PEG 8000 concentrations simultaneously, with fixed imidazole and NaCl concentrations at pH 8.4.

The enzyme crystallised over a broad range of concentrations for the parameters varied (Fig. 1). The time required for crystals to appear was dependent on both enzyme and precipitant concentrations, with the first crystals appearing within 24 h. Conditions which generated the most uniform crystals in a reasonable amount of time were chosen to scale up the crystallisation to 1 ml. Batch crystallisation at the 1 ml scale was carried out at 4 mg ml^{-1} enzyme concentration and 24% w/v PEG. At this scale, crystals appeared within 8 weeks and were small plates which were left to grow to a maximum size of 400 μm in length, 200 m in width and approximately 40 μm thick.

The crystallisation at the 1 ml scale was repeated with a second batch of purified epoxide hydrolase under the same conditions used for the first 1 ml crystallisation. The crystallisation in the second batch gave a relatively low yield of needle shaped crystals, and the formation of a precipitate and a skin layer were noted. The presence of contaminants in a protein solution can have negative consequences on the ability to crystallise a protein. It is possible that a small variation in the epoxide hydrolase batch purity caused the poor repeatability of the crystallisation step.

Cross-linking

Crystal cross-linking was carried out with 5% v/v glutaraldehyde. Cross-linking times of varying duration were tested to determine the effect of degree of cross-linking on CLEhC activity. Activity assays were carried out in a convenient 96-well plate format. Approximately the same number of crystals was contained in each well as determined visually by microscope observation. The substrate concentration profiles recorded showed some scatter in the recorded absorbance (Fig. 2) which was attributed to interference by CLEhCs in the light path. The degree of scatter in the data was variable from well to well. Despite this scatter it was observed that CLEhC activity decreased with increasing cross-linking time (Fig. 2).

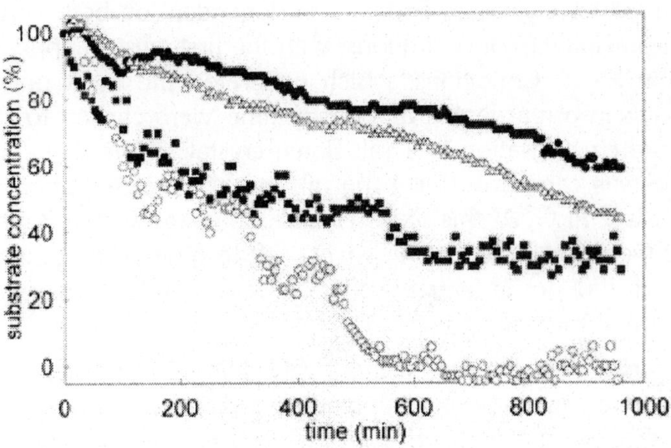

Figure 2: Effect of cross-linking time on CLEhC activity. Hydrolysis of racemic pnitrostyrene oxide (pNSO) at an initial concentration of 150 μM was followed spectrophotometrically at 310 nm. CLEhCs were cross-linked with 5% v/v glutaraldehyde for either (o) 0.5 h, (■) 1 h or (Δ) 24 h. Conversion of pNSO by spontaneous hydrolysis only is shown for comparison (•).

When crystals were cross-linked for 24 h, conversion of pNSO proceeded at a rate equal to spontaneous pNSO hydrolysis, indicating that the CLEhCs were completely inactive. Crystals which were cross-linked for the shortest time (0.5 h) showed the highest activity. CLEhCs used in all further experiments were cross-linked for 0.5 h. When CLEhCs were stored in TEM buffer at 22°C, no change in the protein concentration of the incubation solution (less than 1 μg ml^{-1} measured by the Bradford method) could be detected over a period of at least 4 months indicating that CLEhCs were resistant to dissolution in buffer.

This was confirmed by testing the buffer for epoxide hydrolase activity after incubation with CLEhCs. The pNSO conversion rate in this liquid was the same as that for spontaneous pNSO hydrolysis, again indicating that no active enzyme leaches from the cross-linked crystals. Our results clearly show that the time of cross-linking is inversely related to the retained activity of the CLEhCs. Costes et al. indicated a trade-off between CLEC stability and activity after cross-linking, so that the degree of cross-linking should be the minimum possible to ensure sufficient stability without loss of activity.

It would be useful to perform a more thorough screening of cross-linking conditions, to compare the effect on CLEC stability and activity of cross-linking carried out using a low concentration of cross-linking agent for a longer time with cross-linking carried out at higher cross-linking agent concentration for shorter times. The optimal degree of cross-linking using glutaraldehyde is likely to vary for different enzymes depending on the number and location of lysine residues contained in the enzyme molecule.

The reactivity of the lysine side chains is determined by their spatial accessibility and their natural interactions with neighbouring side chains. Epoxide hydrolase contains 17 lysine residues which are located mainly on the outer surface of the enzyme molecule, exposed to solvent and far from the active site. It seems unlikely that cross-links directly interfere with substrate conversion at the active site. More long ranging structural effects or lack of enzyme mobility at the active site, however, cannot be excluded.

X-ray structural studies of the wild-type epoxide hydrolase unexpectedly found the Gln134 side chain positioned in the active sites which was explained as a consequence of crystal packing forces and the affinity of the active site residues for the amide group. Other structural effects could arise from crystallisation and the resulting packing forces and would need to be investigated by determining the X-ray structure of the Y215F mutant.

CLEhC Activity

In order to reduce the amount of CLEhCs needed per experiment, an assay was developed for which reasonable changes in substrate concentration could be measured with reaction volumes of 200 µl. A typical progress curve for the hydrolysis of racemic pNSO is shown in Figure 3. The progress curve displays a biphasic shape with a change in turnover rate after 50% of the total substrate was converted. The second phase of conversion had a rate close to that of spontaneous hydrolysis indicating a relatively high enantioselectivity, which is supported by the results reported below.

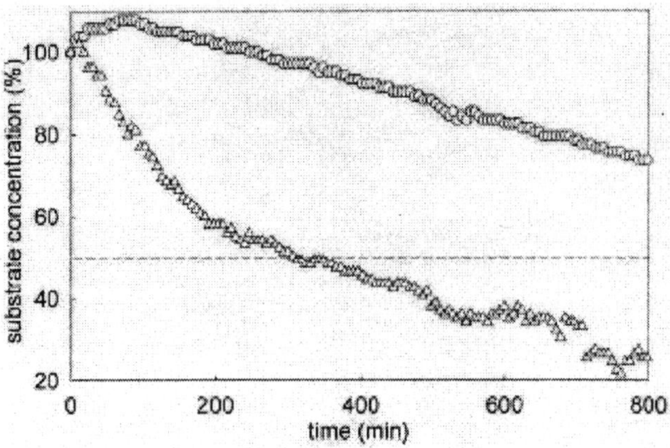

Figure 3: Typical conversion of racemic pNSO in CLEhC activity assays measured spectrophotometrically in the 96 wells-plate arrangement. Initial pNSO concentration was 150 µM in 200 µl TEM buffer. Both (Δ) CLEhC conversion of racemic pNSO and (o) spontaneous hydrolysis of pNSO are shown.

Good repeatability of the assay was essential for carrying out stability experiments, which were based on comparison of activity before and after incubation under certain conditions. The repeatability of the assay was tested for the conversion of 150 μM racemic pNSO using a 96-well plate reader spectrophotometer. Initial activities calculated for successive assays were comparable and ranged within a standard deviation of 20% from the average activity (Table 1).

Table 1. Reuse of CLEhCs for the conversion of racemic pNSO in successive batches.

Assay[a]	Initial rate, CLEhC[b] (dA_{310}/dt)	Initial rate, spont.[c] (dA_{310}/dt)
1	$2.4x10^{-4}$	$4.6\ x10^{-5}$
2	$2.4x10^{-4}$	$4.4\ x10^{-5}$
3	$3.7x10^{-4}$	$4.3\ x10^{-5}$
4	$2.7x10^{-4}$	$3.8\ x10^{-5}$
5	$2.8x10^{-4}$	$5.0\ x10^{-5}$

[a] A batch of CLEhCs was used for the conversion of 150 M racemic pNSO in 200 l TEM buffer, pH 7.5. Substrate conversion was followed by measuring the change in absorbance at 310 nm (A_{310}). Initial rates were calculated for no more than 20 % conversion. The same batch of CLEhCs was used for the five assays and between each assay the CLEhCs were washed 5 times with TEM buffer.
[b] Average initial enzymatic hydrolysis rate (dA_{310}/dt) = $2.8x10^{-4}\pm5.2x\ 10^{-5}$.
[c] Average initial spontaneous hydrolysis rate (dA_{310}/dt) = $4.4x10^{-5}\ \pm0.4x10^{-5}$.

These results also show that the CLEhCs could be used at least 5 times without loss of activity, demonstrating good reusability which is desirable for large-scale application of CLEhCs. The specific activity of the CLEhC preparation was estimated by measuring the initial rate of conversion of pNSO by the spectrophotometric method described above, using a known amount of enzyme in the CLEC form.

The amount of enzyme used for each conversion was determined visually using a polarised light microscope by counting the individual CLEhCs and estimating their size. The concentration of enzyme in the crystals was estimated to be 8 mM. Three separate reactions with different amounts of CLEhCs, visually observed to be approximately in the ratio 1:1.5:2.7, were carried out to obtain confidence in the calculated specific activity.

For the soluble Y215F epoxide hydrolase mutant, the K_m values reported for the (*R*) and (S) enantiomers of pNSO were >0.8 and >1 mM respectively. Based on these values, the reactions were assumed to occur under first order conditions since the pNSO concentration in the assays was low (150 μM). This seems to be supported by the first order conversions observed using an even higher substrate concentration (0.5 mM) as shown in Fig. 7A.

For these calculations the effects of mass transfer limitations were neglected. The observed rates of reaction were roughly proportional to the amount of enzyme present in each reaction volume. Based on the estimated amount of enzyme used for the conversions, an average apparent first order rate constant (k^R_{cat}/K^R_m) of 0.07 0.01 $mM^{-1}s^{-1}$ was calculated. Since for the soluble epoxide hydrolase the specificity constant for the conversion of (S)-pNSO is two orders of magnitude lower than that of (*R*)-pNSO, the measured activity for CLEhCs was assumed to be mostly attributable to the conversion of (*R*)-pNSO. Therefore, the determined first order rate constant was taken as the specificity constant (k^R_{cat}/K^R_m) (*R*)-pNSO.

Based on this specificity constant, the activity of the CLEhC preparation was approximately 2% of that of the soluble Y215F epoxide hydrolase. The residual activity of CLECs has been reported to range from just a few percent less, to less than 1% of the activity of the soluble enzyme. The specific activity is affected by size of the crystals and the relative size of the substrate and the crystal channels. If mass transfer limitations were present, then the specific activity determined by this method would be an underestimation of the intrinsic activity of the enzyme in the crystals. A change in the conformation of the enzyme in the CLEC form can also cause a decrease in specific activity as discussed below.

Enantioselectivity

The enantioselectivity of the CLEhCs was tested by performing kinetic resolutions of racemic pNSO and racemic styrene oxide (SO) (Fig. 4, and Fig. 5). The observed enantioselectivity was compared to that of the free enzyme in solution. The enantiomeric ratio, E, for the resolution of the racemic epoxides was calculated by fitting the degree of conversion, X, and the enantiomeric excess of the epoxide substrate, ees, to the well known formula (Fig. 6).

$$E = \frac{\ln[(1-X)(1-ee_S)]}{\ln[(1-X)(1+ee_S)]}. \tag{1}$$

The apparent enantioselectivity of a kinetic resolution can be reduced as a result of the occurrence of unwanted side reactions. n this case, nonenantioselective spontaneous epoxide hydrolysis takes place during kinetic resolution. Because this side reaction causes a decrease in the yield of remaining enantiopure epoxide enantiomer, the E-ratio determined using the Chen formula is an underestimation of the intrinsic E-ratio for the resolution.

The first order chemical hydrolysis rate constants for pNSO and SO were determined under the same reaction conditions as the enzyme reaction but in the absence of enzyme. The E-ratio corrected for spontaneous hydrolysis was determined by fitting the progress curves (Fig. 4A and 5A) to competitive Michaelis-Menten kinetics taking into account chemical hydrolysis. The fitted kinetic constants were used to calculate the E-value by Eq. 2. The corrected E values, especially for styrene oxide, were still substantially lower than those found for the soluble enzyme.

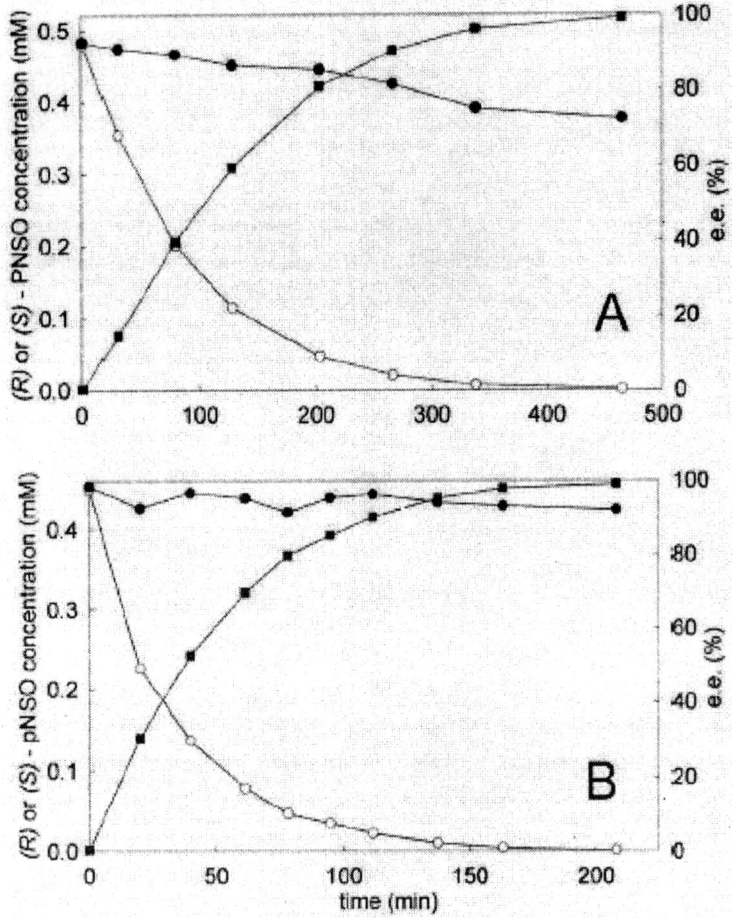

Figure 4: Kinetic resolution of 1 mM racemic pNSO by: ***A****, Y215F CLEhCs,* ***B****, soluble Y215F epoxide hydrolase. (•) (S)-enantiomer, (O) (R)-enantiomer, (■) e.e.*

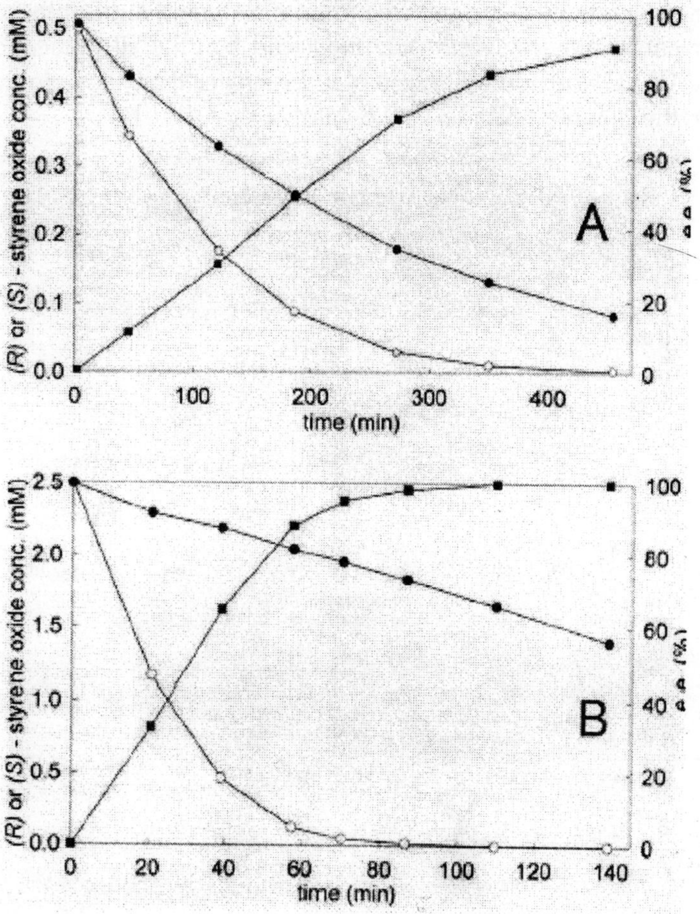

Figure 5: Kinetic resolution of: ***A****, 1 mM racemic styrene oxide by Y215F CLEhCs,* ***B****, 5 mM racemic styrene oxide by soluble Y215F epoxide hydrolase. (•) (S)-enantiomer, (o) (R)-enantiomer, (■) e.e.*

The enantioselectivity was also determined by comparing the rate of conversion of the individual enantiomers of the epoxides at the same CLEhC concentration. All single enantiomer conversions were carried out at 0.5 mM substrate concentration (Fig. 7). It was assumed that the conversions could be described by first order kinetics. This assumption was validated by the good first order exponential fit obtained for each data set, which also suggested that the K_m values for the CLEC epoxide hydrolase may be higher than those found for the soluble enzyme (>0.8 mM).

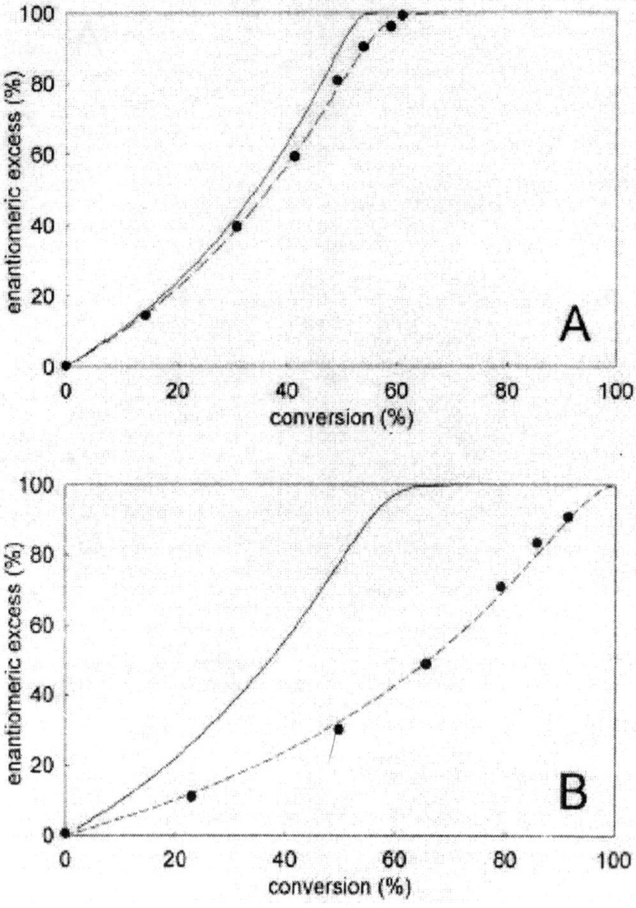

Figure 6: Enantioselectivity of epoxide kinetic resolutions carried out using soluble or CLEhC Y215F epoxide hydrolase. ***A****: Resolution of 1 mM racemic pNSO.* ***B****: Resolution of 1 mM (using CLEhCs) or 5 mM (using soluble enzyme) racemic styrene oxide.*

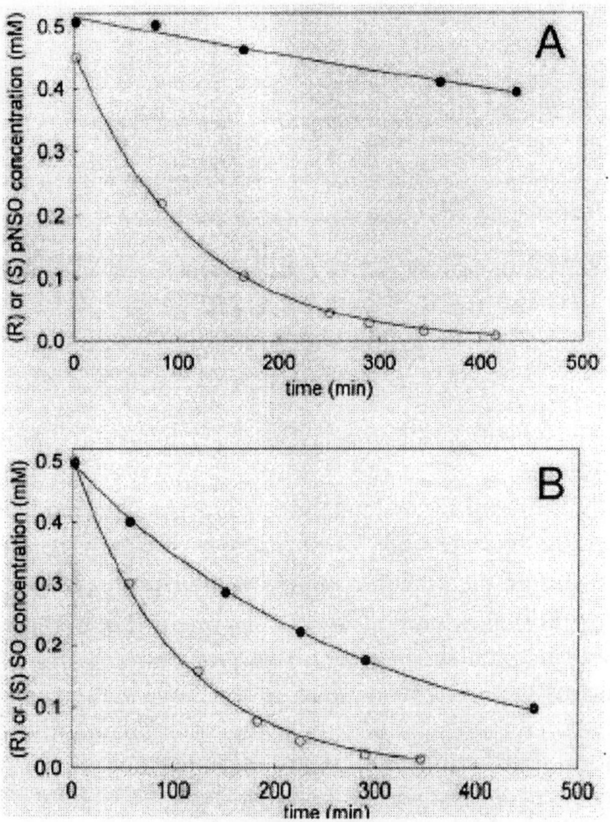

Figure 7: Conversion of epoxide pure enantiomers by CLEhCs carried out separately. ***A****: Enzymatic hydrolysis of 0.5 mM (o) (R) and (•) (S)-pNSO.* ***B****: Enzymatic hydrolysis of 0.5 mM (o) (R) and (•) (S)-styrene oxide.*

The E-values were calculated by fitting the progress curves (Fig. 7) to first order kinetics taking into account spontaneous hydrolysis. The enantiomeric ratio was then determined according to,

$$E = \frac{(k_{cat}/K_M)_R}{(k_{cat}/K_M)_S} \tag{2}$$

The values obtained were comparable to those determined from the kinetic resolution of the racemic mixtures. In both cases, the apparent enantiomeric ratio found using the CLEhCs was considerably lower than that found using the soluble enzyme.

Mass Transfer Limitations in CLEhCs

The apparent enantioselectivity of heterogeneous catalysts can be reduced as a result of mass transfer limitations. Mass transfer limitations cause the formation of intra-particle concentration gradients for each enantiomer. The gradients are more severe for the faster reacting enantiomer, effectively reducing the availability of this enantiomer to the catalyst. The low apparent enantioselectivity calculated from the progress curves suggests that either mass transfer limitations in the CLEhC preparation are considerable, or that CLEhCs have different conversion kinetics compared to the soluble enzyme form.

For the following considerations, first order kinetics were assumed since the substrate concentrations used in the CLEhC conversions were low compared to the known Km values for conversion of these epoxides with the soluble Y215F mutant. The good first-order fits of the progress curves for the conversion of the individual enantiomers by CLEhCs (Fig. 7) support this assumption. The rate of conversion of either enantiomer is described by,

$$\frac{dS}{dt} = -k_{obs}^{S} S + k_c S \,, \tag{3}$$

where, k_c is the first order rate constant for spontaneous chemical hydrolysis of epoxide. The observed first order rate constant for enzymatic conversion of the (S)-epoxide, k^S_{obs} is given by,

$$k_{obs}^{S} = \frac{k_{cat}^{S}}{K_{M}^{S}} \eta_{S}[E], \tag{4}$$

and similarly, for the (*R*)- epoxide,

$$k_{obs}^{R} = \frac{k_{cat}^{R}}{K_{M}^{R}} \eta_{R}[E] \tag{5}$$

where, k_{cat} and K_m are the intrinsic Michaelis-Menten kinetic constants, [E] is the enzyme concentration and is the effectiveness factor. The effectiveness factors, S and R, represent the reduction in reaction rate due to mass transfer limitations. When there is no mass transfer limitation, then $\eta_S = \eta_R = 1$. By combining equations 2, 4 and 5, the following well-known relation between the intrinsic E value, E, and the effective (observed) E value, E_{eff}, can be derived,

$$E_{eff} = E \frac{\eta_R}{\eta_S}. \tag{6}$$

If mass transfer limitations exist and the enzyme concentration is constant, then $\eta_R < \eta_S$ since R is the faster reacting enantiomer. The effectiveness factor is a function of the Thiele modulus (Φ) which represents the relationship between the maximum rate of reaction and the maximum rate of mass transfer according to,

$$\Phi = \frac{\delta}{2} \sqrt{\frac{k_{cat}[E]}{K_m D_{eff}}} \tag{7}$$

where, δ is the thickness of the crystals, and D_{eff} is the effective diffusivity of the epoxide molecules in the CLEhC pores. For a first order reaction, and slab geometry,

$$\eta = \frac{\tanh \Phi}{\Phi} \tag{8}$$

From equations 6, 7 and 8 it follows that when $\Phi \rightarrow \infty$ (mass transfer limited system) $E_{eff} \rightarrow \sqrt{E}$.

The degree of mass transfer limitations which occur in the CLEhCs can be established by estimating the Thiele modulus for the faster reacting (*R*)-enantiomer, Φ_R. Generally, mass transfer limitations occur when $\Phi > 0.3$, which from Eq. 8 corresponds to $\eta = 0.97$. Clearly, to estimate the Thiele modulus the enzyme concentration in the crystals is required.

It proved difficult to determine [E] experimentally, but an estimate can be derived from knowledge of the crystalline structure of the wild-type epoxide hydrolase from X-ray crystallographic studies. The study determined that a molecule of epoxide hydrolase has dimensions of 48 x 47 x 47 $Å^3$ in the crystal lattice and that the solvent content of the crystal was 52%. Accordingly, the enzyme concentration in the CLEhCs can be calculated as 8 mol cm^{-3} (8 mM), which is in good agreement with other estimates made of enzyme concentration in the crystalline form.

The maximum thickness, δ, of the CLEhCs was visually estimated to be 40 μm. The effective diffusivity was estimated based on the molecular diffusivity of styrene oxide in water (estimated as 5 x 10^{-6} cm^2 s^{-1}) and taking into account effects of steric hindrance based on the relative size of the CLEhC pore size (25 Å) and the molecular diameter of styrene oxide, estimated as 7 Å. Accordingly, D_{eff} was calculated as 6 x 10^{-7} cm^2 s-1. In the literature values of D_{eff} have been assumed to be approximately 50% of the free solution values, D_m, therefore our estimate of D_{eff} is a conservative one.

Based on the estimated specificity constant k^R_{cat} / K^R_m of 0.07 $mM^{-1}s^{-1}$ it follows that Φ_R = 1.98 and therefore η_R = 0.48. Assuming that, similar to the soluble enzyme, the specific activity for the conversion of (*S*)-pNSO for the CLEhCs is 2 orders of magnitude lower than for (*R*)-pNSO, η_S = 0.99.

From Eq. 6, with E=87 determined for the soluble enzyme, the corresponding value of E_{eff} for pNSO would be 42. This value is still higher than that observed experimentally for the conversion of pNSO by CLEhCs. This implies that the decrease in enantioselectivity cannot be fully explained by the occurrence of internal mass transfer limitations.

The calculated values of are approximations since all the parameters used for calculations (δ, k^R_{cat} / K^R_m, [E], D_{eff}) are estimates. For unchanged specific activity and for equal [E] and Deff, mass transfer limitations would not exist if the width of the CLEhC was reduced by a factor of 10, to less than 4 µm. From the calculations outlined above, they conclude that the decrease in enantioselectivity observed for kinetic resolutions using CLEhCs was at least partially caused by internal mass transfer limitations.

External diffusion limitations, which arise due to a stagnant liquid layer around the CLECs can also reduce the observed rate of reaction, when S $< K^S_m$ or R $< K^R_m$, and enantioselectivity. The stagnant liquid layer can be eliminated or reduced by sufficient mixing of the reaction medium so that external mass transfer limitations become negligible. It is assumed that external diffusion limitations were greatly reduced during the kinetic resolution experiments since these were well mixed.

However, conversions carried out in the 96-well plates, which were shaken intermittently, may have suffered from such limitations even though increasing the frequency of agitation did not increase the rate of reaction. The existence of external mass transfer limitations in the activity measurements carried out in the microwells would reduce the specific activity determined for the CLECs. This would result in an underestimation of the Thiele modulus, which was based on this measured value of specific activity, and would therefore also result in an underestimation of the decrease in enantioselectivity caused by the internal mass transfer limitations.

The possibility that the reduced enantioselectivity observed with the CLECs was at least partially caused by a change in the intrinsic conversion kinetics of the Y215F epoxide hydrolase mutant in the CLEC form cannot be ruled out. Enzymatic catalytic efficiency, k_{cat}, has often been found to be lower in the crystalline form, whereas both increases and decreases in the substrate binding affinity, K_m, have been reported.

These changes have been attributed to minor structural changes that can occur in the enzyme active site upon cross-linking or crystallisation due to distortion of the crystal lattice, and due to restrictions in the mobility of the active site during conversion. It has been suggested that co-crystallisation with an inhibitor of the enzyme may help in retaining the correct alignment of catalytic residues in the active site during crystallisation, thereby retaining enzyme activity and enantioselectivity.

It is possible that co-cross-linking with a competitive inhibitor may also be used to the same effect. A small protective effect against loss of activity afforded by a competitive inhibitor during crosslinking of carboxypeptidase-A with glutaraldehyde was demonstrated by Quiocho and Richards.

Product Inhibition

The soluble form of the Y215F epoxide hydrolase mutant, which has been crystallised in this work, shows competitive inhibition by various diols to a similar extent as the wild-type enzyme. To test further the suitability of the CLEhC preparation for large-scale use, the degree of inhibition by two diols, 1-phenyl-1,2-ethanediol (PED) and 3-phenoxy-1,2-propanediol (PPD), the products of styrene oxide and pNSO hydrolysis, respectively, was investigated.

The competitive inhibition constants (Ki) for the soluble Y215F mutant are 31 mM and 49 mM for (S) and (*R*)-PED respectively, and 6 mM for racemic PPD. Inhibition of the CLEhC preparation by these diols was tested by measuring the enzymatic activity for the conversion of racemic pNSO by CLEhCs incubated in diol solutions at various concentrations, compared to their activity in buffer. Results showed that no significant decrease in activity occurred when the CLEhCs were incubated in the diol solutions (Table 2).

Table 2: Inhibition of CLEhCs by (R) or (S)-1-phenyl-1,2-ethanediol (PED) and 3-phenoxy-1,2-propanediol (PPD).

Diol	Diol conc. (mM)	Change %
	25	+ 11
(*R*)-PED	50	- 9
	75	+ 2
	100	+ 27
	15	+ 34
(*S*)-PED	30	+ 7
	52	+ 14
	75	+ 22
	3.1	+ 3
PPD	6.2	- 21
	9.3	+ 21
	12.4	+ 5

The decrease in activity expected in the presence of diol at the concentrations tested and with such low substrate concentration (150 μM) would be much larger than the 20% standard deviation in the activity measurements. This suggested that the inhibition constants for these diols were much higher for the CLEhCs than for the soluble enzyme, indicating reduced inhibition. A reduced inhibition of the crystalline form of enzymes compared to the soluble forms has been reported in several different cases.

For example, a 100-fold reduction in inhibition of carboxypeptidase-A crystals by β-phenylpropionic acid was found compared to the soluble enzyme. The lower apparent inhibition could be caused by a change in binding affinity of the diols in the CLEhCs active site, or by a reduced partitioning of the diol in the crystal lattice, which is expected to be a relatively hydrophobic environment compared to the aqueous buffer liquid.

Indeed, the application of CLECs as chromatographic separation material is partly based on hydrophobic interactions between the intra-CLEC environment and the compounds to be separated. Size exclusion effects are generally also important but are unlikely to have much effect in this case since the size of the diol molecule is expected to be much smaller than typical crystal pore sizes. The reduced inhibition effect of diols on the CLEhCs would be an advantage in the use of CLEhCs for large-scale kinetic resolutions.

Stability

The thermostability of the CLEhC preparation was tested by comparing activity of a batch of CLEhCs before and after incubation at 60°C for different incubation periods. At 60°C the CLEhCs appeared to be more thermostable than the soluble enzyme (Fig. 8). After incubation at this temperature for 1 h the CLEhCs retained their activity completely, whereas the activity of the soluble enzyme decreased by 20% in the same time.

The decrease in CLEhC activity seemed to level off after 100 min at 60% of the original activity, but after 22 h at 60°C CLEhC activity had decreased by 70%. The decrease in CLEhC activity was correlated with an irreversible loss of crystal birefringence, which was observed visually under a polarised light microscope, although no change in the macroscopic shape of the crystals was observed. A similar loss of crystal birefringence has been observed in other studies concerned with thermostability of protein crystals, and has been attributed to structural changes of the protein within the crystalline environment.

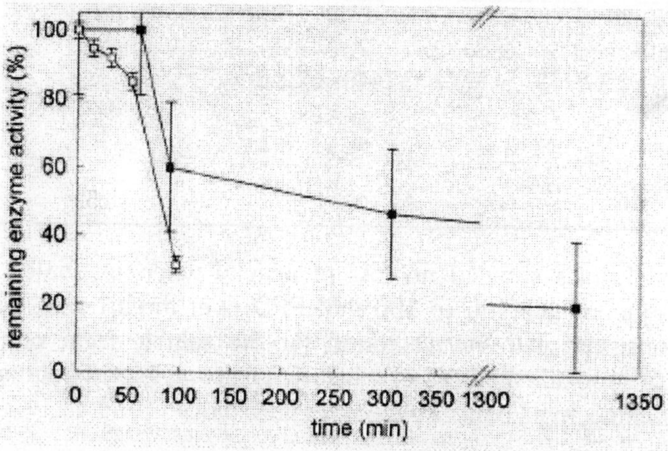

Figure 8: Thermostability of (□) soluble Y215F epoxide hydrolase and (■) Y215F CLEhCs. Enzyme preparations were incubated at 60°C and remaining enzyme

In large-scale processes, the use of high epoxide concentrations is desirable to obtain high reaction productivity. Cosolvents can be used to increase the concentration of epoxides in aqueous solutions since epoxides are often poorly water soluble. They investigated the stability of the CLEhCs in solutions of DMSO to determine whether CLEhCs could be used in the presence of cosolvents.

When incubated with 50% v/v DMSO, the CLEhCs proved to be considerably more stable than the soluble enzyme form (Table 3). However, a decrease in CLEhC activity in time was observed, with 30% of the original activity left after

24 h, indicating that only a lower DMSO concentration would allow sustained enzyme activity and CLEhC reuse.

Table 3: Stability of CLEhCs and soluble Y215F epoxide hydrolase in 50% DMSO

Enzyme form	Cumulative incubation time (min)	Remaining activity (%)
	5	100
CLEhC	70	67
	1440	33
Soluble	5	100
Y215F	42	49
	100	25

The CLEhCs showed no loss of activity when incubated in an aqueous solution of 100 mM PPD over three hours. This is consistent with results found for the soluble form of the Y215F epoxide hydrolase, which was also stable in solutions containing high concentrations of this diol.

Soluble Thermostability

When the thermostability of the soluble Y215F epoxide hydrolase mutant was measured for comparison to the CLEhC preparation it was found that this mutant in the soluble form showed unexpectedly high stability at 60°C. Earlier studies have shown that this active site mutant is more stable than the wild-type enzyme at high pH and also when incubated in solutions of diols, such as (PED and PPD) at high concentration.

At 50°C, the wild-type enzyme underwent a much more rapid decrease in activity than the mutant. The activity of the wild-type enzyme decreased by 50% after 65 min incubation whereas the same reduction for the mutant occurred only after 1390 min. At 60°C, however, the mutant activity decreased relatively rapidly and 50% activity loss occurred after approximately 80 min.

Chapter 13

Properties of Thermostable Enzymes

Thermostable enzymes, which have been isolated mainly from thermophilic organisms, have found a number of commercial applications because of their overall inherent stability. Advances in this area have been possible with the isolation of a large number of beneficial thermophilic microorganisms from different exotic ecological zones of the earth and the subsequent extraction of useful enzymes from them.

While the most widely used thermostable enzymes are the amylases in the starch industry, a number of other applications are in various stages of development. In the food related industry, they have been used in the synthesis of amino acids. In the petroleum, chemical and pulp and paper industries, for example, thermostable enzymes have been used for the elimination of sulphur containing pollutants through the biodegradation of compounds like dibenzothiophene, in the production of 1,3-propanediol from glycerol and in replacing polluting chemical reagents causing toxic products.

Currently, a number of publications have extensively discussed developments in this area. Adaptation of extremophiles to hot environments, production of heat-stable enzymes from thermophiles and hyperthermophiles, structure and function relationships of thermozymes (heat-tolerant enzymes) and biotechnological and industrial applications of thermostable enzymes are among the topics that have been studied.

In situ temperatures between 80 and 115°C were found to be conducive biotopes for a number of hyperthermophiles. Some examples are mentioned in Table 1. Extremophiles are also adapted to live at low temperatures in the cold polar regions, at high pressure in the deep sea and at a very low and high pH values (pH 0-3 or 10-12), or at a very high (5-30%) salt concentration.

Table 1: Examples of sites serving as sources of microorganisms which can provide thermo tolerant enzymes

Source	Microorganism	Enzyme
Hot spring	Thermus sp.	α-Amylase
Hot spring	Bacillus sp. WN.11	α-Amylase
Deep sea hydro-thermal vent	Staphilothermus marinus	α-Amylase
Marine solfatare	Thermococcus litoralis	Pullulanase
Decomposed plant samples	Clostridium absonum	Cellulase free xylanase
from a lake	CFR-702	
Hot spring	Bacillus thermoleovo-cans ID-1	Lipase
Compost of fermen-ting citrus peels, coffee and tea extract	Bacillus strain MH-1	Endochitinase
residues		
Compost Bacillus stearothermophilus	CH-4	β-N-cetylhexosaminidase
Korean salt fermented anchovy	Bacillus sp. KYJ963	β-Amylase
Deep sea hydrothe-rmal vent	Pyrococcus abyssi	Alkaline phosphatase
Sediments of hot	Bacillus sp. 3183	α-Amylase-like

springs		pullulanase
Garbage dump	Bacillus circulans	Xylanase
Compost treated with artichoke juice	Bacillus sp.	Inulinase

In the last decade, a number of hyperthermophilic archaea, the least understood domain of life have been isolated and are able to grow around the boiling pcint of water. The organisms with the highest growth temperatures (103-110°C) are members of the genera Pyrobaculum, Pyrodictium, Pyrococcus and Methanopyrus. Within the bacteria, Thermotoga maritima and Aquifex pyrophilus exhibit the highest growth temperatures of 90 and 95°C respectively. These properties imply extremely important industrial and biotechnological implications due to the fact that enzymes from such microorganisms can be employed for use in harsh industrial conditions where their specific catalytic activity is retained.

Resistance of Thermophiles

Microorganisms, like all living things, adapt to the condition in which they have to live and survive. Thermophiles are reported to contain proteins which are thermostable and resist denaturation and proteolysis. Specialized proteins known as chaperonins are produced by these organisms, which help, after their denaturation to refold the proteins to their native form and restore their functions.

The cell membrane of thermophiles is made up of saturated fatty acids. The fatty acid provides a hydrophobic environment for the cell and keeps the cell rigid enough to live at elevated temperatures. The archae, which compose most of the hyperthermophiles, have lipids linked with ether on the cell wall. This layer is much more heat resistant than a membrane formed of fatty acids. The DNA of thermophiles contains a reverse DNA gyrase which produces positive super coils in the DNA. This raises the melting point of the DNA (the temperature at which the strands of the double helix separate) to at least as high as the organisms' maximum temperature for growth. Thermophiles also tolerate high temperature by using increased interactions that non-thermotolerant organisms use, namely, electrostatic, disulphide bridge and hydrophobic interactions.

Thermostable enzymes are stable and active at temperatures which are even higher than the optimum temperatures for the growth of the microorganisms. Thermophilic enzymes are more rigid proteins than their mesophilic counterparts. A clearer understanding of this capacity should be possible with new methodologies that clearly indicate changes in protein structure.

While determining the conformational entropy of both enzymes, the folded state showed a higher structural exibility for the thermophilic protein than the mesophilichomologue. It has, thus, been assumed that a mechanism characterized by entropicstabilization could be responsible for the higher thermostability of the thermophilic enzyme.

Advantages of Enzymes from Thermophiles

Thermostable enzymes are gaining wide industrial and biotechnological interest due to the fact that their enzymes are better suited for harsh industrial processes. Some biocatalytic conversions and industrial applications of thermostable enzymes are presented in Table 2. One extremely valuable advantage of conducting biotechnological processes at elevated temperatures is reducing the risk of contamination by common mesophiles.

Allowing a higher operation temperature has also a significant influence on the bioavailability and solubility of organic compounds and thereby provides efficient bioremediation. Other values of elevated process temperatures include higher reaction rates due to a decrease in viscosity and an increase in diffusion coefficient of substrates and higher process yield due to increased solubility of substrates and products and favourable equilibrium displacement in endothermic reactions.

Table 2: Bioconversion reactions and applications of thermostable enzymes

Enzyme	Temperature range (°C)	Bioconversions
α-Amylase (bacterial)	90–100	Starch→dextrose syrups
α-Amylase (fungal)	50–60	Starch→dextrose syrups
Pullulanase	50–60	Starch→dextrose syrups
Xylanase	45–65, 105a	Craft pulp→xylan+lignin
Chitinase	65–75b	Chitin→chitobiose
		Chitin→N-acetyl glucosamine (chitibiase)
		N-acetyl glucosamine→ glucosamine (deacetylation)
		Chitin→chitosan (deacetylase)
Cellulase	45–55, 95c	Cellulose→glucose
Protease	65–85	Protein→amino acids and peptides
Lipase	30–70	Fat removal, hydrolysis, interesterification, alcholysis, aminolysis
DNA polymerase	90–95	DNA amplification

[a] Xylanase from Thermotoga sp.
[b] Within this range enzyme activity was high.
[c] Cellulases from Thermotoga sp.

Such enzymes can also be used as models for the understanding of thermostability and thermo-activity, which is useful for protein engineering. The following are the specific thermostable enzymes, their sources, characteristics and application requirements of high temperature resistance.

Amylolytic Enzymes

The starch industry is one of the largest users of enzymes for the hydrolysis and modification of this useful raw material. The starch polymer, like other such polymers, requires a combination of enzymes for its complete hydrolysis. These include a-amylases, glucoamylases or β-amylases and isoamylases or pullulanases. The enzymes are classified into endo-acting and exo-acting enzymes. α-amylase is an endo-acting enzyme and hydrolyses linkages in a random fashion and leads to the formation of linear and branched oligosaccharides, while the rest are exo-acting enzymes and attack the substrate from the non-reducing end, producing oligo and/or monosaccharides.

The starch hydrolytic enzymes comprise 30% of the world's enzyme consumption. The enzymatic conversion of all starch includes gelatinization, which involves the dissolution of starch granules thereby forming a viscous suspension, liquefaction, which involves partial hydrolysis and loss in viscosity, and saccharification, involving the production of glucose and maltose via further hydrolysis. Gelatinisation is achieved by heating starch with water, and starch is water-soluble only at high temperatures which are dependent on the source.

For hydrolysis of the starch to proceed immediately after gelatinization, hence, among other things avoiding a lot of cooling time, the enzyme has to be thermostable. A number of attempts were made to isolate and characterize amylolytic enzymes from diversified sources to meet the requirements of the starch industry. Table 3 shows thermostable microbial starch hydrolyzing enzymes and their properties. Thermostable a-amylases were isolated a long time ago from Bacillus subtilis, Bacillus amyloliquefaciens and Bacillus licheniformis.

Table 3. Source microorganisms and properties of thermostable starch hydrolyzing enzymes

Enzymes	Organism	Enzyme properties	
		Optimal temp.(°C)	Optimal pH
a-Amylase	Bacillus amyloliquefaciens	70	7.0
	Bacillus licheniformis	100	6.0–6.5
	Bacillus stearothermophilus	70–80	5.0–6.0
	Bacillus stearothermophilus	70	–
	Bacillus subtilis	70	7.0
	Lactobacillus manihotivorans	55	5.5
	Myceliophthora thermophila	100	5.6
	Pyrococcus furiosus	100	5.5
	Pyrococcus woesei	100	6.5–7.5
	Staphylothermus marinus	65	5.0
	Sulfolobus solfataricus	–	–
	Thermococcus aggreganes	100	5.5
	Thermococcus celer	90	5.5
	Thermococcus fumicolans	95	4.0–6.3
	Thermococcus hydrothermalis	85	4.8–7.8
	Thermomyces lanuginosus	60	5.6
	Thermococcus profoundus	80	4.0–5.0
β-Amylase	Bacillus circulans	60	–
	Bacillus cereus var. mycoides	50	–
	Bacillus sp.	50	7.5

Clostridium thermosul-phurogenes	75	5.5
Clostridium thermo-sulfurogenes	60	5.8–6.0
Pullulanase Bacillus sp.	60	–
Pyrococcus furiosus	98	5.5
Pyrococcus woesi	100	5.5–6.0
Thermococcus aggregans	100	6.5
Thermus caldophilus GK	24	75 5.5
Thermococcus celer	90	5.5
Thermococcus hydro-thermalis	95	5.5
Thermococcus litoralis	98	5.5
Thermotoga maritima MSB8	90	6.0

The commercially available β-amylases with a catalytic activity up to 60°C were isolated from various Bacillus species. They can be used in conjunction with debranching enzymes to produce pure maltose syrups. The pullulanases that were produced in earlier work on different microorganisms were not very suitable for operation under the conditions prevailing in the industry. However, many thermophilic microorganisms have since been found to produce pullulanases

Termamyl and Fungamyl are two well-known amylolyic enzymes which are now available commercially. These enzymes are used world-wide for the production of glucose syrups and syrups with different level of dextrose equivalent. Hetero-oligosaccharides have been synthesized by Termamyl obtained from B. licheniformis. In the presence of soluble starch and added non-starch sugars, oligosaccharides were produced by the reversed catalytic reaction. One of the concerns of the starch industry is the calcium requirement of the a-amylase enzymes and the formation of calcium oxalate, a substance that may block process pipes and heat exchangers.

Besides, such accumulation in some products like beer is not acceptable. Its precipitation can be reduced through a decrease in calcium ion requirement of enzymes and lowering of the pH of the production process. Activities of a number of the starch hydrolyzing enzymes, however, are calcium dependent although the degree of utilization varies (Table 4). Production of calcium independent and acid stable a-amylases has been attempted by protein engineering and as a result Termamyl LCe, which was produced via site-directed mutagenesis, has shown high calcium independence.

Table 4: Calcium requirement of industrially important starch degrading enzyme

Enzyme	Micro-organism	Application temperature range (°C)	Application pH range dosage	Minimum Ca^{2+} (ppm)
Bacterial mesophilic α-amylase	Bacillus subtilis	80–85	6.0–7.0	150
Bacterial thermophilic α-amylase	Bacillus licheniformis	95–105	6.0–7.0	20
Fungal α-amylase	Aspergillus oryzae	55–70	4.0–5.0	50
Amyloglucosidase	Aspergillus niger	55–65	3.5–5.0	0
Pullulanase	Bacillus acidopolluliticus	55–65	3.5–5.0	0

Although, raw starch dominates in nature there are few enzymes that can catalyze its hydrolysis efficiently. Saccharification of liquefied starch is carried out at low pH values. However, currently used thermostable a-amylases are not stable at such low pH. An economical process could be attained through the use of amylases stable at the saccharification stage. In addition to this, a one-step process of starch hydrolysis using amylolytic enzymes would decrease the costs of glucose production. The anticipated diversity of species of thermophiles within high temperature environments and the requirements of the new and improved technological operations for enzymes with novel and fitting characteristics have placed a challenge both for the scientific and business community.

Thermostable Xylanases

Xylan, which is the dominating component of hemicelluloses, is one of the most abundant organic substances on earth. It has a great application in the pulp and paper industry. The wood used for the production of the pulp is treated at high temperature and basic pH, which implies that the enzymatic procedures require proteins exhibiting a high thermostability and activity in a broad pH range. Treatment with xylanase at elevated temperatures disrupts the cell wall structure. This, as a result, facilitates lignin removal in the various stages of bleaching.

Xylanases for such a purpose (1) must lack cellulytic activity to avoid hydrolysis of the cellulose fibers, (2) need to be of low molecular mass to facilitate their diffusion in the pulp fibers, and (3) most importantly, high yields of enzyme must be obtained at a very low cost. All commercially available xylanases can only partially fulfill these requirements, and the optimum temperature for the activity of most xylanases is reported to be 50-60°C with a half-life of about 1 h at 55°C However, some xylanases have been reported to exhibit higher thermal stability and optimal activity ranging from 80 to 100°C.

These enzymes have offered a major step in the reduction of chlorine consumption in the bleaching process of kraft pulp, thus, lowering environmental pollution by organich alogens. The enzymes are generally classified in to two families. Family 10 groups (EXs 10) having high molecular weights (>30 Kda) and Family 11 (EXs 11) which are low molecular weight (<30 Kda) xylanases. Among the two families, the EX 10-fold is found to be more thermostable. Due to their capacity to enhance the bleaching of kraft paper, however, EXs 11 have remained more attractive.

Thermostable xylanase were isolated from a number of bacterial and fungal sources (Table 5). Members of the Bacillus sp., Streptomyces sp., Thermoascus aurantiacus and Fusarium proliferatum have been reported to produce xylanases which are active at temperatures between 50 and 80°C. While the Dictyoglomus sp. were described to produce xylanases operating at an optimum temperature of 90°C, a number of Thermotogales sp. were reported to secrete thermostable xylanases which can function at higher temperatures.

Table 5: Source microorganisms and properties of thermostable xylanases

Organism	Enzyme properties	
	Optimal temp.(°C)	Optimal pH
Baccillus amyloliquefaciens	80	6.8–7.0
Bacillus circulans	80	6.0–7.0
Bacillus sp.	60–75	8.0–9.0
Bacillus sp. strain SPS-0	75	6.0
Bacillus subtilis	50	6.0
Clostridium abosum	75	8.5
Dictyoglomus sp. strain B1	90	6.0–7.0
Fusarium proliferatum	55	5.0–5.5
Pyrococcus furiosus	100	6.0
Pyrococcus furiosus	102	–
Scytalidium thermophilum	65	6.0
Streptomyces sp. strain S38	60	6.0
Sulfolobus solfataricus	105	5.3
Teheromyces lanuginosus (wild and mutant)	60, 70	7.0, 6.7
Teheromyces lanuginosus-SSBP	70–75	6.5
Thermoascus aurantiacus	50	5.0
Thermotoga maritima MSB8	92	6.2
Thermotoga maritima MSB8	95	6.0–7.5

Thermotoga maritima MSB8	75	6.2
Thermotoga neapolitana	95	6.0
Thermotoga neapolitana	85	5.5
Thermotoga neapolitana	102	5.5
Thermotoga sp. strain FjSS3-B1	80	7.0
Thermotoga sp. strain FjSS3-B1	105	6.8–7.8
Thermotoga sp. strain FjSS3-B1	115	5.3
Thermotoga thermarum	80	6.6

Alkaliphilic and cellulase-free xylanases with an optimum temperature of 65°C from Thermoactinomyces thalophilus subgroup C and cellulase-free xylanases from Clostridium abusonum CFR-702 were also reported recently. From one Thermotoga sp. an enzyme active at 115°C was produced and characterized. It was also possible to produce xylanases in solid-state fermentation, where a decreased cultivation time and an increased concentration of basal medium improved its production. This is a promising aspect from an economicpoint of view.

The pulp and paper technology is one of the fastest growing industries and the use of thermostable xylanases seems attractive since they provide global environmental benefits. However, scaling up of the enzyme production from the respective microorganisms to the level required by the industry remains to be seen. It is also worth mentioning that, extreme thermophiles that are able to secrete xylanase are few. The search for a thermophile with high yield of enzyme and the desired characteristics is still being pursued.

Thermostable Cellulases

Cellulose, the most abundant organic source of feed, fuel and chemicals consists of glucose units linked by b-1,4-glycosidic bonds in a linear mode. The difference in the type of bond and the highly ordered crystalline form of the compound between starch and cellulose make cellulose more resistant to digest and hydrolyze. The enzymes required for the hydrolysis of cellulose include endoglucanases, exoglucanases and β-glucosidases. While cellulase is an endoglucanase that hydrolyses cellulose randomly, producing oligosacaccharides, cellobiose and glucose, exoglucanases hydrolyze β-1,4-D-glucosidic linkages in cellulose releasing cellobiose from the nonreducing end. On the other hand, β-glycosidases of thermophilic origin, which have received renewed attention in the pharmaceutical industry hydrolyze cellobiose to glucose.

In the current industrial processes, cellulolytic enzymes are employed in the colour extractions of juices, in detergents causing color brightening and softening, in the biostoning of jeans, in the pretreatment of biomass that contains cellulose to improve nutritional quality of forage and in the pretreatment of industrial wastes. In order to attack the native crystalline cellulose, which is water insoluble and occurs as fibers of densely packed structures, however, thermostable cellulases active at high temperature and high pH are required.

Thermostable cellulases of archaeal origin include those isolated from Pyrococcus furiousus and Pyrococcus horikoshi. While the latter has an optimum temperature of 97°C, the enzyme from Pyrocuccus furiousus has shown optimal activity at 102-105°C (Table 6). Sulfolobus solfataricus MT4, Sulfolobus acidocaldarius and Sulfolobus shibatae were also described as producers of b-gucosidases. From Thermotoga maritema MSB8 optimally active cellulase acting at 95°C and between pH 6.0 and 7.0 was reported.

Other endocellulases (CelA and CelB) from Thermotoga neapolitana, were purified and characterized. Optimal pH for CelA is 6.0 at 95°C, and the pH range of CelB is broad (6.0-6.6) at 106°C. CelA, with the ability to hydrolyze microcrystalline cellulose, was isolated from the extremely thermophilicbac terium Anaerocellum thermophilum and maximal activity of this enzyme was observed at pH 5.0-6.0 and 85-95°C.

Table 6: Source microorganisms and properties of thermostable cellulases

Organism	Enzyme properties	
	Optimal temp.(°C)	Optimal pH
Anaerocellu thermophilum	85–90	5.0–6.6
Bacillus subtilis	65–70	5.0–6.5
Pyrococcus furiosus	102–105	–
Pyrococcus horicoshi	97	–
Rhodothermus marinus	95	6.5–8.0
Thermotoga maritema MSB8	95	6.0–7.0
Thermotoga neapoltana (EndocellulaseA)	95	6.0
Thermotoga neapoltana (EndocellulaseB)	106	6.0–6.6

It is obvious that, the successful utilization of cellulose is dependent on the development of economically feasible technologies for the production of cellulase. The biopolishing process of cotton in the textile industry, for example, requires cellulase stable at high temperature close to 100°C. Presently used enzymes for this purpose, however, are active only at 50- 55°C. Cellulase production is also found to be the most expensive step during ethanol production from cellulosic biomass, and accounted for approximately 40% of the total cost. In the food industry, degradation of cellulose by acids is still unsatisfactory and results in the decomposition of the sugars. Finally, even though many cellulolytic enzymes of thermophillic origin are known their function under physiological condition remains unclear.

Thermoactive Chitinases

Chitin, the second most abundant natural biopolymer after cellulose, consists of a linear β-1,4-homopolymer of N-acetylglucoseamine residues. In nature, it is usually found attached to other polysaccharides and proteins. The covering layer of insects, cell walls of various fungi and crab and shrimp wastes are the main sources of chitin. The major applications of chitosan include wastewater clearing, preparation of cosmetics, paper production, textile finishes, photographic products, cements, heavy metal chelating agents and for medical and veterinary purposes.

The production of chitosan from chitin involves a deacetylation reaction that is done using 40% sodium hydroxide solution. This is a highly corrosive stream which is difficult to control. Attempts are thus being made to do this conversion using deacetylase enzyme. Perhaps a thermostable deacetylase enzyme will help in increasing yields and conversion rates and allow reuse of the enzyme.

Endo-acting chitin hydrolase chitinaseA, the exoacting hydrolase chitinaseB and N-acetyl-D-glucoseaminidase are responsible for chitin degradation. However, chitin is not easily accessible to chitinases and chitin deacetylases. Evidences from X-ray diffraction studies have shown that chitin is a highly ordered crystalline structure and does not dissolve in water. This property of chitin has shown the need for thermostable enzymes.

The thermophilic organisms Bacillus licheniformis X-7u, Bacillus sp. BG-11 and Strteptomyces thermoviolaceus OPC-520 were reported to be the major sources of chitinases. Chitinase and Nacetyl-glucosaminidase were produced from the extreme thermophilican aerobic arachaeon Thermococcus chitinophagus. A thermophilicbac - terium strain producing three different endochitinases in its culture fluid was isolated from chitin-containing compost. The strains belonged to the genus Bacillus and three isoforms of endochitinases (L, M and S) were purified and characterized.

They had different molecular masses and showed temperature optima of 75, 65 and 75°C at a pH of 6.5, 5.5 and 5.5 respectively. The iso-electric points were 5.3, 4.8, and 4.7. Thermostable exochitinases were also isolated from

Bacillus stearothermophilus CH-4, isolated from a compost of organic solid wastes.

The asicparts of shrimp and prawns used for consumption are the muscle tissues. The resulting waste of the products, thus, is vast. In some developing countries it was reported that more than 2.5 million tones of shrimp and prawns are produced per year. Therefore, there is a high need to develop methods of producing value added products from this waste. As the major way to exploit the potential from this waste is the production of chitin and its derivative chitosan, improved and economically feasible chemical, enzymatic and novel biotechnological conversion bioprocesses are required. Chitin together with its derivatives can also be used as agrochemicals. However, there is only little e.ort to clarify its utilization in the food industry.

Thermostable Proteases

Proteases, which are generally classified into two categories are becoming major industrial enzymes, and constitute more than 65% of the world market. These enzymes are extensively used in the food, pharmaceutical, leather and textile industries. The applications will keep increasing in the future as will the need for stable biocatalysts capable of withstanding harsh conditions of operation.

Relative ease of isolation of Bacilli from diverse sources has made these organisms the focus of attention in biotechnology. So far, however, few thermophilic Bacillus sp. that produce proteases have been isolated, the earliest isolate being Bacillus stearothermophilus which is stable at 60°C. Another Bacillus sp. has produced a thermostable protease that has an optimum activity at 60°C, while a di.erent Bacillus stearothermophilus sp. produced an alkaline and thermostable protease which is optimally active at 85°C.

A species of Bacillus stearothermophilus TP26 that has been isolated produces an extra cellular protease having an optimum temperature of 75°C. Enhancement of protease activity excreted from Bacillus stearothermophilus had also been possible using economical chemical additives in the proteolysis reactions involved in waste activated sludge. In a chemically de.ned medium, thermophilic and alkaliphilic Bacillus sp. JB-99 was also reported to produce thermostable alkaline proteases.

Dominant producers of proteases in fact, are the microorganisms of the genera Pyrococcus, Thermococcus and Staphylothermus (Table 7). Extremely thermostable serine proteases are produced by the hyperthermophilicarc haeum Desulfurococcus strain, and thermostable metalloproteases are reported from a gram-negative thermophilic bacterium. Major area of focus in the future concerning the production of proteases is the optimization of media.

Table 7: Source microorganisms and properties of thermostable proteolytic enzymes

Organism	Enzyme properties	
	Optimal temp.(°C)	Optimal pH
Bacillus brevis	60	10.5
Bacillus licheniformis	70	9.0
Bacillus stearothermophilus	60	–
Bacillus stearothermophilus	85	–
Bacillus sp. JB-99	80	6–12
Bacillus stearothermophilus TP26	75	–
Bacillus sp. no. AH-101	80	12.0–13.0
Bacillus thermoruber	45	9
Pyrococcus sp. KODI	100	7
Staphylothermus marinus	–	9
Thermoacidophiles (archeal and bacterial origin)	60–70	7.0–8.5
Thermococcus aggreganes	90	7.0

Thermococcus celer	95	7.5
Thermococcus litoralis	85	8.5
Thermotoga maritema	95	9.5

Syntheticmedia have a great advantage over complex media in that consistency of processes and production is enhanced through avoiding the variability of complex substrates. Thus, syntheticmedia provides better control and monitoring, improved product recovery and quality, and simpli.ed puri.cation systems.

It is for this reason that reports on the production of protease enzymes using syntheticmedia are available recently. There is considerable current interest on the exploration of proteases that can catalyze reactions in cold water. This will allow their use in detergents which can be used in normal tap water without the requirement for increasing the temperature of the water. The search for such enzymes is very much a challenge at this time.

Lipases

Lipases of microbial origin are the most versatile enzymes and are known to bring about a range of bioconversion reactions, which includes hydrolysis, interesterification, esterification, alcoholysis, acidolysis and aminolysis. Their unique characteristics include substrate specificity, stereospecificity, regioselectivity and ability to catalyze a heterogeneous reaction at the interface of water soluble and water insoluble systems. The esters produced play a relevant role in the food industry as flavour and aroma constituents.

Whereas long chain methyl and ethyl esters of carboxylic acid moieties provide valuable oleo chemical species that may function as fuel for diesel engines, esters of long chain carboxylic acid and alcohol moieties (waxes) have applications as lubricants and additives in cosmetic formulations. Other applications include the removal of the pitch from pulp produced in the paper industry, for the hydrolysis of milk fat in the dairy industry, removal of non-cellulosic impurities from raw cotton before further processing into dyed and finished products, drug formulations in the pharmaceutical industry and in the removal of subcutaneous fat in the leather industry. A biodiesel was derived from vegetable oils using immobilized Candida antarctica lipase.

Most of the industrial processes in which lipases are employed function at temperatures exceeding 45°C. The enzymes, thus, are required to exhibit an optimum temperature of around 50°C. According to some reports there are fats exhibiting higher melting points and which are able to inhibit enzymaticreactions at a low temperature. Some enzymatic processes for the physical refining of seed oils have four distinct requirements.

These include pH of 5.0 and optimal temperature of around 65°C, adding an enzyme solution and enzyme reaction followed by the separation of the lysophosphatide from the oil at about 75°C. These reactions, therefore, are enhanced through the utilization of thermo-tolerant lipases.

Lipases are of widespread occurrence throughout the earth's flora and fauna. More abundantly, however, they are found in bacteria, fungi and yeasts. Several Bacillus sp. were reported to be the main source of lypolyticenz ymes. While most of these enzymes are active at a temperature of 60°C and pH of 7.0, lipases from Bacillus thermoleovorans and a thermophilic Rhizopus oryzae strain can moderately function at extreme pH and temperature values. The pH and temperature optima for the catalytic activity of some thermostable lipases are given in Table 8.

Reports of thermostable lipases from archaeal origin, however, are very few. Phospholipase A_2 which was secreted from the archaea Pyrococcus horikoshii was involved in crude oil refining. This enzyme was found to optimally react at 95°C and pH of 7.0. Phospholipase A_2 has a better performance in the degumming processes of oil refineries and thereby reduces wastewater problems and running costs.

Table 8: Source microorganisms and properties of thermostable lipases

Organism	Enzyme properties	
	Optimal temp.(°C)	Optimal pH
Bacillus acidocaldariusa (esterase)	70	–
Bacillus sp. RSJ-1	50	8.0–9.0
Bacillus strin J 33[a]	60	8.0
Bacillus stearothermophilusa	68	–
Bacillus thermocatenletusa	60–70	8.0–9.0
Bacillus thermoleovorans ID-1	70–75	7.5
Geobacillus sp.	70	9.0
Pseudomonas sp.	65	9.6
Pseudomonas sp.	90	11.0
Pyrobaculum calidifontis	90	–
Pyrococcus furiosusa (esterase)	100	–
Pyrococcus horikoshii	97	5.6
Pyrococcus horikoshii	95	7.0

[a]Active at water immiscible solvents.

Among the desirable characteristics that commercially important lipases should exhibit, alkali tolerance and thermostability are the main. To meet this end, there is a continuous search for sources of highly active lipolytic enzymes with specific stability to pH, temperature, ionic strength and organic solvents. Although, few lipases exist which are able to operate at 100°C, their half-lives are reported to be short. The other problem associated with the production of lipases and which requires improvements is its secretion at the late stage of growth. Proteases secreted in the mean-time either change the properties of the lipase produced or degrade it. Perhaps it is for this reason that only few lipases can be obtained in industrial quantities. Hence the need for a more stable lipase enzyme produced in large quantities.

Polymerases Chain Reaction

The polymerase chain reaction (PCR) process has led to a huge advance in genetic engineering due to its capacity to amplify DNA. The three successive steps in this process include denaturation or melting of the DNA strand (separation) obtained at a temperature of 90-95°C, renaturation or primer annealing at 55°C followed by synthesis or primer extension at around 75°C. Development in this process has been to a large extent facilitated by the availability of thermostable DNA polymerases, which catalyse the elongation of the primer DNA strand.

In earlier PCR procedures, DNA polymerases which were isolated from Escheria coli were utilized. These enzymes, however, lost their enzymatic activities at elevated temperatures and, thus, adding a new polymerase enzyme after each cycle following the denaturation and primer hybridization steps was necessary. This process made the thermal cycling a time-consuming and costly procedure. Taq polymerase from the bacterium Thermus aquaticus was the first thermostable DNA polymerase characterized.

Repeated exposure to 98°C in a reaction buffer had little effect on the enzyme activity and significant activity remained after exposure to 99°C. Although, Taq polymerase exhibits a 50-30 exonuclease activity, a 30-50 exonuclease activity was not detected. Thus, the base insertion .delity is low as the enzyme is unable to correct misincorporated nucleotides. It had also been possible to determine the error rates of some polymerases in terms of base pairs.

Table 9 shows a variety of thermostable polymerases with 30-50-exonuclease-dependent proof reading activity required from a high fidelity polymerase. Optimization of the PCR procedure can be done by mixing a standard polymerase, such as Taq polymerase with high .delity polymerases. While the PCR process is developing rap:dly through the invention of better instruments, lack of fidelity has remained to be a serious challenge. The have distinct activities in which errors can occur are the rate of phosphodiester bond formation, the binding of dNTP by the polymerase, the rate of pyrophospahte release, contamination after misincorporation and the 30-50-exonuclease proof reading capacity of the enzyme

Table 9: Thermostable DNA polymerases in PCR

Polymerases	Organism	PCR reactions
BstI pol	Bacillus stearothermophilus	HF
Deep Vent	Pol Pyrococcus sp.	GB-D LF
Pfu pol	Pyrococcus furiosus	LF
Pwo pol	Pyrococcus woesei	LF
Taq pol I	Thermus aquaticus	HF
T. pol	Thermus filiformis	NI
T. pol	Thermus flavus	NI
Tth pol	Thermus thermophilus	HF
Tma pol	Thermotoga maritema	LF
Vent pol	Thermococcus litoralis	LF

HF and LF, high and low fidelity; NI, not identified.

Some issues that nowadays have stimulated scientists in the area thus, are improvements required in the fidelity of the PCR, development of strategies for more economical use of the enzymes and ensuring rapid multiplication from fewer strands. A thermostable DNA polymerase having these characteristics will indeed improve the results obtained by PCR machine.

Molecular cloning of thermophilic genes into mesophilic hosts Genetic and protein engineering are the modern techniques for the commercial production of enzymes of improved stability to high temperatures, extremes of pH, oxidizing agents and organic solvents. Cloning and expression of genomic infor mation available in a thermophile in a suitable and faster growing mesophilichost has also provided possibilities of producing the specifictherm ostable enzyme required for a particular biotransformation process.

Genes encoding α-amylase (amyA), pullulanase (pulA), maltodextrin phosphrylase (agpA) and α-glucosidase (aglA) were identified in the gene library of Thermotoga maritema. The enzymes produced after expression in Escheria coli are reported to demonstrate extreme thermostability. Pyrococus furiosus a-amylase gene, which was cloned by activity screening in Escheria coli was reported to be optimally active at 100°C, pH 5.5-6.0 and not to require Ca^{2+} for its activity.

Genes encoding pullulanases from Pyrococus furiosus and Ferividobacterium pennavorans, Thermotoga neapolitana genes encoding for xylanases, Protease genes from Pyrobaculum aerophilum and Bacillus stearothermophilus, lipases from Bacillus thermocatenulatus, esterases from Bacillus licheniformis and cellulase genes from the archaon Pyrococus furiosus were expressed from an Escheria coli recombinant strain. The production of cellulase enzyme was further enhanced through DNA recombinant technology by encoding the genes for cellulase in various species of Clostridium.

In addition to this, the nucleotide sequence and the cloning of CelA and CelB genes from C. thermocellum were studied and reported. Cloning and expression of chitinase genes was also successfully done. In cases where a relevant gene of a thermostable enzyme has been identified and once the gene has been transferred into a mesophilichost the resultant enzyme production can be facilitated through appropriate incubation processes.

Chapter 14

Isolation of Enzymes from Extremophiles

Extremozymes have a great economic potential in many industrial processes, including agricultural, chemical and pharmaceutical applications. Many consumer products will increasingly benefit from the addition or exploitation of extremozymes. The characterisation of microorganisms that are able to thrive in extreme enviroments has received a great deal of attention: such extremophiles are a valuable source of novel enzymes.

Extreme conditions can refer to physical extremes (e.g. temperature, pressure or radiation), but also to geochemical geochemical extremes such as salinity and pH . Most of the extremophiles that have been identified to date belong to the domain of the Archaea. However, many extremophiles from the eubacterial and eukaryotic kingdoms have also been recently identified and characterised.

The notion that extremophiles are capable of surviving under non-standard conditions in non-conventional environments has led to the assumption that the properties of their enzymes have been optimized for these conditions. Indeed, data for a considerable fraction of the enzymes that have been isolated and functionally characterized from extremophiles support this assumption.

The classification of "extreme environments' refers to a wide variety of different conditions to which microorganisms have adapted. The biocatalysts obtained from these microorganisms could be applicable in similarly diverse conditions. For the degradation of polymers such as chitin, cellulose or starch, enzymes that are active at and resistant to high temperatures are often preferred. Under these conditions the solubility and, consequently, the accessibility of the substrate is improved.

Alternatively, if one needs to perform a stereospecific modification of a compound for the synthesis of a pharmaceutically relevant product in organic solvents, very different prerequisites apply to the biocatalysts. As salt is known to reduce water activity, enzymes from halophilic microorganisms could be the most suitable choice for application in nonaqueous media. This diversity of environments to which different extremophiles have adapted offers many exciting opportunities for a variety of applications.

Thermophilic Extremophiles

Thermophilic extremophiles have attracted most attention. In particular extremophilic proteases, lipases and polymer-degrading enzymes, such as cellulases, chitinases and amylases have found their way into industrial applications. The reasons to exploit enzymes that are stable and active at elevated temperatures are obvious. At elevated temperatures the solubility of many reaction components, in particular polymeric substrates, is significantly improved. Moreover, the risk of contamination, leading to undesired complications, is reduced at higher temperatures.

The structural features of thermophilic extremozymes has attracted much attention. Several three-dimensional structures have been solved and compared with those of mesophilic counterparts, with the ultimate goal of elucidating the mechanisms underlying thermostability. In a comprehensive study, 10 thermophilic and hyperthermophilic serine hydroxymethyltransferases were compared with 53 mesophilic homologs. Structural alignment and homology modeling was applied to identify the different mechanisms involved in thermal stability. For this enzyme class, it was concluded that stability was achieved by a combination of increased surface charge, increased protein core hydrophobicity and replacement of exposed "thermolabile' amino acids.

Psychrophiles

More recently, enzymes from psychrophiles have become interesting for industrial application, partly because of ongoing efforts to decrease energy consumption. For example, there is an increasing desire to apply psychrophilic enzymes in detergents. With such enzymes it becomes feasible to develop laundry applications that can be performed at lower temperatures. For such processes, psychrophilic proteases, amylases or lipases have great commercial potential. The pulp and paper industry is also interested in polymer-degrading enzymes that are active at lower temperatures. Several food processing applications would also benefit from the availability of low temperature enzymes.

A characteristic feature of many enzymes from psychrophiles is the correlation of high catalytic activity and low thermal stability at moderate temperatures, which can be partly explained by the increased flexibility of the molecule, compared with mesophilic and thermophilic enzymes. This adaptation of the enzymes to low temperatures has been the subject of several studies. It has been assumed that increased flexibility is correlated with decreased stability and a delicate balance between stability and activity is indeed often observed.

Nevertheless, there is an increasing number of examples from nature as well as from protein engineering studies showing that the structural features involved in stability or activity can be very different and act independently. A good overview of present knowledge on the properties of enzymes from psychrophiles and their applications can be found in two recent reviews.

Halophiles

Halophiles can survive in hypersaline habitats by their ability to maintain osmotic balance. They accumulate salts such as sodium or potassium chloride (NaCl or KCl), up to concentrations that are isotonic with the environment. As a result, proteins from halophiles have to cope with very high salt concentrations (e.g. KCl concentrations of ~ 4 M and NaCl concentrations of >5 M). The enzymes have adapted to this environmental pressure by acquiring a relatively large number of negatively charged amino acid residues on their surfaces to prevent precipitation. Consequently, in surroundings with lower salt concentrations the solubility of halophilic enzymes is often very poor, which could limit their applicability.

However, this property has been taken advantage of by applying halophilic enzymes in aqueous/organic and nonaqueous media. For example, an extracellular protease from Halobacterium halobium has been exploited for efficient peptide synthesis in water/N0-N0-dimethylformamide. Recently, a p-nitrophenylphosphate phosphatase (p-NPPase) from Halobacterium salinarum was used in an organic medium at very low salt concentrations after entrapping the enzyme in reversed micelles.

Under these conditions p-NPPase was active and stable. Similar observations were made with a halophilic malate dehydrogenase. Exploitation of reversed micelles in combination with halophilic enzymes is likely to result in the development of novel applications for these enzymes.

Alkaliphiles/acidophiles

Enzymes from microorganisms that can survive under extreme pH could be particularly useful for applications under highly acidic or highly alkaline reaction conditions, for example in the production of detergents. However, one of the striking properties of acidophilic and alkaliphilic microorganisms is their ability to maintain a neutral pH internally, and so the intracellular enzymes from these microorganisms do not need to be adapted to extreme growth conditions. However, this does not account for extracellular proteins, which have to function in low or high pH environments in the case of acidophiles and alkaliphiles, respectively.

Proteases, amylases, lipases and other enzymes that are resistant to and active at high pH and high chelator concentrations of modern detergents are desirable. This has prompted the screening of alkaliphilic bacteria and Archaea for their ability to produce such enzymes. By these means, several useful enzymes have already been identified and obtained. Combinations of homologybased PCR and activity screening have been applied to screen for and detect alkaline proteases in a collection of thermoacidophilic archaeal and bacterial strains isolated from hot environments .

In an alternative approach, alkaliphilic bacilli that could grow at >pH 9 were used as a source for oxidation-resistant alkaline proteases . Polymer-hydrolysis related processes have also initiated the search for biocatalysts from acidophiles. Several enzymes used for starch-hydrolysis (e.g. amylases, pullulanases, glucoamylases and glucosidases that are active at low pH) have been isolated.

Piezophiles

It is thought that pressure does not exert a major selective force on protein function . This assumption is based on the consideration that pressures exceeding 400 MPa are needed to induce the denaturation of single-chain proteins. In view of that, it should be noted that even microorganisms that live in the deep sea are not exposed to pressures that exceed 120 MPa .

Under those conditions, their enzymes do not need specific pressure-related adaptations. However, there are some examples of the specific stabilization of proteins by increased pressure. Pressure-resistant proteins could be of use, in particular for food production, where high pressure is applied for processing and the sterilization of food materials . The biotechnological opportunities for piezophiles have recently been reviewed in detail .

At present, it is clear that no matter how extreme the conditions are at defined locations on Earth there is a fair chance that microbes will be able to survive. Additional examples further to those described above are microorganisms that grow in the presence of high metal concentrations (metallophiles), at high radiation levels (radiophiles), or under oxygen deprivation (microaerophiles).

The biotechnological application of enzymes from such extremophiles is not always obvious. Nevertheless, in view of the great potential of biocatalysis it is very likely that new concepts will be developed that will result in the application of enzymes from these and other extremophiles in industrial processes.

Extreme Genomes

The growing number of genomes available from extremophiles will greatly aid the discovery and identification of useful novel enzymes. Currently, at least 10 genomes from Archaea have been entirely sequenced and another 12 are close to completion . In addition, a significant number of genomes from extremophilic bacteria have and are currently being elucidated. Illustrative of the large potential of extremophiles as a source for novel enzymes is the observation that was made after determining the 1.56 Mb genome of Thermoplasma acidophilum .

In this genome, approximately 1500 open reading frames (ORFs) were identified. For 240 of these ORFs no recognizable homologs could be found in DNA databases available in the public domain. It is tempting to assume that a considerable fraction of the proteins encoded by these "unknown' ORFs is responsible for the unique ability of this microorganism to thrive under extreme conditions. In addition, among this group of "unknown' genes there could be a considerable amount of genes that encode enzymes that are of industrial interest.

This growing wealth of data will therefore be useful for the identification of enzymes that have not been detected by functional screening procedures. This is illustrated by the observation that within the genomes of archaeal Methanosarcina spp multiple homologs for enzymes involved in methanogenesis were present . Expression of the individual genes was shown to be strongly dependent on the composition of the growth medium. Thus, functional screening could have prevented the identification of all potentially useful gene products.

Enzymes Production with Extremophiles

Functional activity screening and mining of DNA data can be very helpful for the identification of useful enzymes. However, the identification of potentially useful enzymes is only a first step. To take full advantage of this knowledge, the identified enzymes have to be made available for in-depth assessment of the biocatalytic properties and application tests. There are two strategies for the production of enzymes from extremophiles. First, production of the biocatalysts can be optimized by increasing the biomass production of the extremophile. Alternatively, the gene encoding the biocatalyst can be cloned and expressed in a suitable host.

Extremophile Biomass Production

Specialized equipment is often a prerequisite to obtain sufficient amounts of a specific biocatalyst directly from the extremophile. In addition, tailor-made fermentation processes will have to be developed and many issues will have to be addressed in the cultivation of extremophiles. Biomass production can be improved either by optimization of the medium composition or by varying the specific fermentation procedures (e.g. fed-batch, cell recycling or continuous cultivation).

In addition, special bioreactors such as gas-lift fermenters can and often have to be used. For example, specific corrosion-resistant bioreactors had to be developed in the cultivation of extreme halophiles for the production of polyglutamic acid (PGA) . Other specific requirements that often have to be met are related to the high pressures that are characteristic of many environments from which the extremophiles are isolated such as the deep sea .

Cloning and Overproduction

At present, the fermentation or production technology for extremophiles is not considered to be standard procedure. Therefore, other ways of making useful biocatalysts available from these sources have to be developed. The cloning and expression of encoding genes in mesophilic hosts can be an attractive alternative method. A major portion of the extremozymes that have found their way into industrial applications are produced using standard Escherichia coli expression systems.

Genetic tools for E. coli expression are very well developed and many plasmid vectors and inducible gene expression systems are available. In particular, the production of thermophilic enzymes is facilitated by the efficient denaturation of most E. coli proteins following a short heat treatment. However, the well-known phenomenon of inclusion-body formation by overexpressed protein in E. coli, could force one to search for alternative hosts.

Fortunately, there is an increasing number of alternative hosts available, including bacterial systems such as Bacillus, Pseudomonas, Lactobacillus, Lactococcus and eukaryotic systems such as Pichia, Kluyveromyces, Candida and Hansenula. The objectives of the expression should be kept in mind in selecting a specific heterologous host system. For example, if the enzyme needs glycosylation for biological activity the selection of a bacterial host should be avoided, whereas for the production of an extracellular enzyme a bacterial host such as Bacillus is likely to be the obvious choice.

A drawback of most Bacillus-based systems is the relatively high level of endogenous protease production by the host, which often leads to degradation of the heterologous gene product of interest. Recently, an extracellular protease-deficient B. brevis system was developed that results in better production levels of heterologous proteins .

Several of the expression systems mentioned above are applicable for the construction of genome libraries. In general, E. coli has been the host of choice for these applications. Nevertheless, in view of the risk of inclusion-body formation and the fact that E. coli has poor protein export capabilities, alternatives should also be considered. The availability of good cloning and expression systems can also be used for improving and tailoring biocatalytic properties of extremozymes by protein engineering and random mutagenesis.

Chapter 15

Predictive Metabolic Engineering

The goal of plant metabolic systems biology is to create a comprehensive multidimensional representation of all of the biosynthetic reactions in a plant. However, this model must also indicate metabolic fluxes: the amounts of intermediates and products that accumulate in a specific tissue or cell type and in each organelle at a given moment in time, and the sizes of the "arrows" or the rates at which the enzymatic and nonenzymatic conversions take place.

This futuristic virtual wall chart must be dynamic, with a time dimension showing the responses of each metabolite pool and flux to changes in environment (light, pathogen, cold, etc.) and developmental state. Finally, to truly qualify as systems biology, the descriptions need to exist in a computer model from which we can make specific testable hypotheses; for instance, that overexpression of a newly discovered tree fern cytochrome P-450 in specific apical cells would increase auxin biosynthesis in these cells and alter the morphology of an ornamental plant in a desirable way.

This level of sophistication demands a complete description of all pathways in at least one plant, the most obvious candidate being Arabidopsis. However, to be comprehensive and useful for the basic and applied biology communities, we will need to build upon this virtual metabolic network with information about pathways that only exist in specific species. Examples would include pathways of secondary metabolism, central metabolic processes that differ across the plant kingdom (C4 versus C3 metabolism, for instance), or storage of carbon in specialised organs or in chemical forms distinct from our canonical "virtual plant" (e.g. starch in tubers).

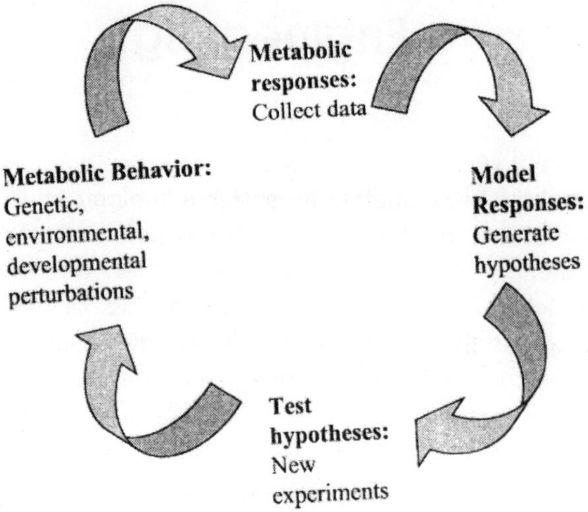

Figure 1: Model building and data accumulation are a cyclical process.

As ambitious as it seems, it will not be enough to have a complete understanding of the biogenesis of small molecules in plants and the changes in these pathways that occur during development and in re-sponse to the environment. To create a model that is predictive, it will be critical to understand the enzymes of these pathways: when and where they accumulate in the cell, how their activity is modulated by specific covalent modifications, how their turnover is regulated, and how their tertiary structures and higher order interactions modulate their functions.

For each of these levels of regulation, we will need comprehensive information about signal transduction pathways, transcriptional regulation, protein trafficking, and more. This means that we cannot do metabolomics in isolation: We need to analyze the data in concert with other large and complex sets of information such as proteomic, global mRNA expression and large-scale structural biological data.

Anaylsis of such complex datasets is a challenging task that requires the adoption of standardised data nomenclatures and formats and the application of sophisticated mathematical and statistical approaches to mine meaningful information from a sea of numbers. Daunting as the task is, a proper integration of transcriptomic, proteomic, and metabolomic datasets will provide the opportunity to analyse metabolic networks at an unprecedented level of detail and will afford a new system-wide level of understanding of the regulation of plant metabolism. Indeed, this approach has already been successful in providing new information about several areas of plant metabolism including specific branches of secondary metabolism, the response to nitrate, and the response to diurnal changes.

Methods in Small Molecule Analysis

The lofty goal of a predictive model for all of plant metabolism requires accurate and reproducible measurement of all metabolites in the plant. A major impediment to this goal is that there is a tradeoff intrinsic to current analytical technologies: We can either accurately measure a relatively small number of molecules in an assay, or qualitatively analyse larger numbers of compounds. Thus, as a rule, with increasing metabolite coverage, there is a loss of accuracy in detection and quantification, which is schematically represented in Fig. 2.

In the last few years, plant metabolite analysis has shifted from specific enzyme assays and non-coupled chromatographic separations that provide information on single compounds or on mixtures of low complexity toward methods offering both high accuracy and sensitivity in highly complex mixtures of compounds (Fig. 2). The approaches of choice for large-scale metabolome analysis are predominantly reliant on coupled mass spectrometric methodologies. For example, gas chromatography-mass spectroscopy (GC-MS) technologies allow the detection, identification, and robust quantification of a few hundred metabolites within a single extract.

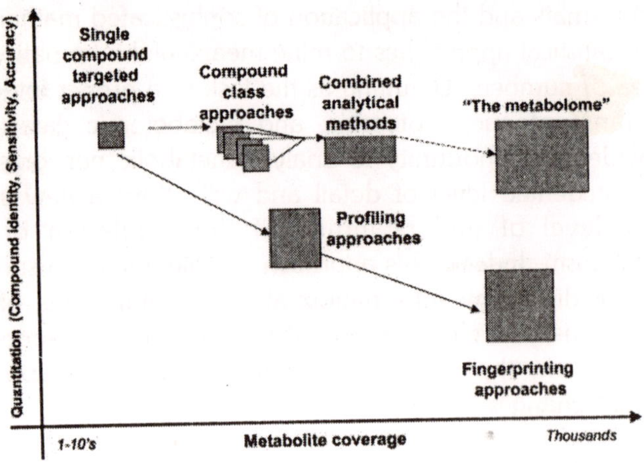

Figure 2: Schematic representation of the different approaches to metabolite measurements

Furthermore, liquid chromatography-MS (LC-MS)-based methodologies have recently been established that allow reproducible determination of several important classes of secondary metabolites including alkaloids, flavonoids, glucosinolates, isoprenes, oxylipins, phenylpropanoids, pigments, and saponins. However, no currently available LC-based method allows the accurate determination of all of these compounds from a single extract, partly due to differences in solubility and stability, and the enormous concentration range of small molecules. LC-MS also suffers from so-called matrix effects, where the ability of the spectrometer to detect and accurately quantify any given ion is influenced by the presence of other ions in the mixture.

The fact remains that the vast majority of metabolites are not measured by approaches currently in use because the sum total of these methodologies covers hundreds rather than the thousands of metabolites estimated to be present in plant cells. Small-molecule analysis therefore currently represents the least comprehensive of the various levels of system analysis, with the current proportional coverage of the profiling technologies decreasing following the order mRNA > protein > metabolite. That said, as is the case in the field of proteomics, progress in analytical chemistry is driving rapid advances in metabolomics.

Analytical tools are currently being developed for the multiparallel analysis of metabolites include time-of-flight-MS (TOF-MS), Fourier transformation-MS (FT-MS), FT-infrared spectroscopy, and NMR spectroscopy, with each technique having comparative advantages and disadvantages and selection of a given tool being largely driven by the biological question at hand. Generally speaking, the choice of method must be determined by evaluating the metabolite coverage it affords with its relative accuracy of compound identification and quantification.

Of these tools, TOF-MS serves a very similar purpose to GC-MS with the exception that TOF mass analyzers are preferred to the quadrupole mass analyzers of standard GC-MS machines, because they can provide a combination of high-mass accuracy and extended range of metabolites detected. Of course, GC-MS is only useful for volatile compounds or molecules that can be rendered volatile by derivitization, whereas LC-MS can be directly adapted to a wider array of molecules.

Nevertheless, GC-MS still has major advantages over LC-MS methods. First, it is easier to achieve complete ionization of molecules in a complex mixture from samples coming off of a gas chromatograph, minimizing interference between molecules in the MS detector. Second, compound identification is greatly facilitated by availability of extensive databases of fragmentation patterns. Finally, tools exist for high-throughput capture of GC-MS data with less time-consuming human intervention than is needed for LC-based methods.

Fourier Transform Ion Cyclotron Mass Spectrometry (FT-IC-MS)

This technique is wholly reliant on the presence of a mass analyzer of sufficient accuracy to allow the definitive empirical formulae for several hundred ions. As with all technologies, disadvantages exist in the use of FT-MS; the most problematic of these is the fact that the lack of chromatography renders the technique unable to differentiate between isomers because of their identical molecular masses. Therefore, metabolites with biologically distinct functions such as Glc and Fru cannot be distinguished from one another.

Furthermore, given the limited amount of published data and the lack of documentation of vigorous method validation, it is unclear how useful this method will be when used in isolation. However the coupling of this or any of the other technologies mentioned above with further fragmentation gives far greater accuracy of identification, but it comes with a heavy cost attached in terms of sample throughput.

The use of FT-infrared spectroscopy and NMR has promise but thus far has not been applied to the quantification of large numbers of compounds in a complex mixture. These methods are well known for identification and quantitation of known molecules, especially when pure or in relatively simple mixtures. They are more recently being tested for metabolic fingerprinting in which the presence and intensity of specific spectroscopic signatures are correlated with molecules of interest in the hope of identifying major changes in metabolite levels or in accurate determination of certain chemicals. NMR has the further advantage that it can be used noninvasively and can yield (albeit limited) spatial resolution of solute concentrations.

Even the parallel use of the existing techniques will only manage to capture a proportion of the metabolome. However, the speed of recent advances in separation sciences and analytical instrumentation is unprecedented. This combined with the opportunity to couple various separation and analytical machines together make it likely that quantum leaps will soon be made toward the goal of accurate measurement of thousands of molecules from complex biological samples.

Cellular and Subcellular Resolution

Because nearly all plant tissues are composed of many cell types, the development and refinement of techniques allowing metabolite analysis of the contents of single cells or subcellular compartments are of great importance. Single-cell sampling techniques now combine cell biological methodologies with physical sampling techniques, and the recent development of high-precision laser capture and catapulting methodologies will most likely produce rapid advances in this area.

In addition to these methods, metabolite-linked bioluminescence assays have been established for visualizing metabolite contents in tissue slices, albeit for a limited number of metabolites at low sensitivity. Thus, at the single-cell level, resolution of metabolite levels is approaching that at the protein and RNA levels.

Obtaining metabolite information at subcellular levels remains particularly problematic. Unlike proteins with their targeting sequences, the location of a metabolite cannot be inferred from their structures. Furthermore, most metabolic intermediates turn over too rapidly to allow measurement after the fractionation procedures regularly used in protein analysis. Despite these problems, several methods have been developed to obtain subcellular information on metabolite levels in intact plants-two of which appear particularly promising.

The first of these, nonaqueous fractionation of lyophilized material, involves the separation of cellular compartments on an organic density gradient and then uses simultaneous equations to estimate metabolite concentrations with respect to marker enzymes of various organelles in the cell. This approach allows the discrimination of the vacuole, plastid, and cytosol and has been used for a wide variety of plant tissues. However, the inability to distinguish the mitochondrion from the cytosol and the apoplast from the vacuole and the fact that this technique requires in excess of 3 g of tissue limit its utility.

The second promising method relies on the generation of specific chimeric proteins created by the fusion of periplasmic binding proteins to green fluorescent proteins that are differentially fluorescent after binding of a given metabolite. Although when taken together, these methods offer great potential, it will be challenging to modify these methods for high-throughput analyses.

Measurement of Metabolic Fluxes

Although measurements of steady-state levels of metabolites give a useful snapshot of the metabolic network at a given moment in time, the true behaviour of plant metabolism can only be gained by direct measurement of metabolic fluxes. Ideally, we need to be able to observe the dynamics of metabolism as it happens. Despite the fundamental importance of metabolic flux, measurement of flux remains a poor cousin to the omic relations of transcripts, proteins, and even metabolites. The basic approach of following the distribution of a labeled precursor through metabolism has been used for many years to estimate fluxes through the major pathways of primary metabolism, but relatively crude methods of determining label distribution have severely limited the number of fluxes that can be measured.

The recent emergence of more sophisticated approaches, based on a combination of steady-state stable-isotope labeling and NMR or MS-based detection systems, may revolutionize our ability to measure flux. This approach allows the determination of positional labeling of a range of metabolites. The position of labeling within the carbon skeleton of end products can allow retrospective evaluation of the metabolic route by which they were formed This technique has already been successfully used to investigate the metabolism of lipids and proteins and has the potential to measure a much greater range of specific fluxes, including those for pathways that operate in parallel in different subcellular compartments.

Nevertheless, the interpretation of stable-isotope-labeling experiments is not a trivial task, and mathematical frameworks must be established that allow the calculation of pathway fluxes from fractional enrichments of isotopic label in each intermediate. In this respect, the plant community is still some way off from the comprehensive flux frameworks that have been developed for microbial organisms; in part, this can be put down to the increased complexity of the plant metabolic network conferred by multiple cell types and extensive subcellular compartmentation.

Modeling Plant Metabolism

Science in its purest form is the reduction of the behaviour of a system to a set of mathematical rules that define it. Ideally, such rules should be sufficiently simple that they themselves provide new insight into system behaviour. Constructing models of metabolism has the potential to achieve this insight and is also an excellent way of succinctly representing large data sets. As such, models can be viewed as the mathematical engine of the virtual wall chart. In addition, models also allow the characteristics of the system to be systematically tested in a way that is too time consuming to achieve experimentally and are thus an important source of new hypotheses.

Models are also extremely useful in pinpointing those parameters that are the most important in determining a particular function, parameters that need to be the focus of greater experimental precision. Historically, metabolic models have concentrated on pathway flux due to the availability of theoretical approaches that can be used to derive flux control structures. Most of the models of plant metabolism to-date are based around the theorems of metabolic control analysis and derive the control-distribution of pathway flux from the kinetic constants of the enzymes involved.

Although these kinetic models have been of some use in understanding the control of plant metabolic pathways, particularly the photosynthetic Calvin cycle, their reliance on difficult-to-acquire experimental information means that they have not been used more widely in metabolism and remain somewhat limited in scope. Stoichiometric analysis is an interesting alternative approach that has gained much attention recently. The basis of this approach is to define elementary flux modes-non-decomposable subnetworks that account for every possible flux within the network. This allows one to mathematically define and describe all metabolic routes that are both stoichiometrically and thermodynamically feasible and is an extremely useful tool for the definition of network structure.

Although flux models provide a description of metabolic dynamics, for these models to enable predictive metabolic engineering, they must also incorporate experimental information about the changes in gene expression and protein abundances that underlie these fluxes (Fig. 1). Ultimately, one would wish to be able to predict exactly which genes should be altered in expression to generate specific changes in particular fluxes. However, bringing together such disparate data sets presents a considerable challenge. A particular problem is the issue of ontology. Put simply, there are too many different ways of naming, describing, and conceptualizing biological elements, leading to description of the same element in several different ways.

Efforts are being made to establish a standard vocabulary that can be applied across all levels of the system hierarchy within an organism and between different organisms. Such ontologies not only furnish standardized descriptors for genes and proteins, but also provide an interactive hierarchical map of the entire system from the level of cellular component, to molecular function, to biological process. An additional problem when integrating different types of data is that the data set becomes multidimensional.

Development of new mathematical approaches that can cope with increased dimensionality will be crucial if data are to be interpreted at a systems level. Machine learning provides a promising approach to this problem. This method uses a variety of different algorithms to provide simple rules that map back onto the measured variables and thereby provide explanations about the behaviour of the system. An alternative approach is to establish relevance networks in which mutual information is used to link data points that follow the same pattern of change.

Although such methods are extremely powerful, they are not readily approachable by those without specialist knowledge. If such techniques are to become a routine part of the analysis of genomic data sets, then software interfaces will be required that allow the user to input and define their data while the algorithms are applied automatically and the results displayed in a readily interpretable format. The display of multidimensional data brings its own challenges, and there are a number of emerging solutions to this problem, examples of which include data touring and the use of glyphs-visual objects onto which many different data attributes are mapped using different visual attributes such as size and colour.

Despite these difficulties in attempting to integrate global data sets, the rewards for successfully doing do so are clear, and several recent papers focusing on subsets of plant metabolic pathways have provided data sets comprising transcripts, enzyme activities, and metabolites. These include studies of various aspects of carbon metabolism including the responses to nitrate and to diurnal changes and studies of individual branches of secondary metabolism such as triterpene saponin biosynthesis. Even though these studies are restricted in their coverage of metabolic space, they nicely demonstrate the power that combining genomic techniques brings to network understanding and pathway elucidation. Expansion of such approaches, even using the tools already at hand, has vast potential in aiding the understanding of complex change underlying diverse biological patterns such as developmental processes or circadian rhythms.

Bibliography

Bommarius, A.S., Riebel, B., *Biocatalysis*, Wiley-VCh, Weinheim, (2004)

Bruggink, A., (Ed.), *Synthesis of β-lactam antibiotics*, Kluwer Acad. Publ., Dordrecht, (2001)

Collins, A.N., Sheldrake, G.N., Crosby, J., (Eds.), *Chirality in Industry,* J. Wiley & Sons, Chichester, (1992)

Godfrey, T., Reichelt, J., (Eds.), *Industrial Enzymology*, 2nd Ed., pp. 435–482, Stockton Press, New York, (1996)

Kirst, H.A., Yeh, W.-K., Zmijewski, Jr., M.J., (Eds.), *Enzyme Technologies for Pharmaceutical and Biotechnological Applications.*, Marcel Dekker, New York, (2001)

Marcel Dekker, New York, Straathof, A., Adlercreutz, P., (Eds.), *Applied Biocatalysis*, Harwood Academic Publishers, Amsterdam, (2000)

McLaren, A.D., Packer, L., Some aspects of enzyme reactions in heterogeneous systems, *Adv. Enzymol, 33*, 245–303, (1970)

Mosbach, K., (Ed.), *Immobilized enzymes*, Methods Enzymol, Vol. 44, Academic Press, New York (1976)

Roberts, S.M., Turner, N.J., Willets, A.J. and Turner, M.K., *Biocatalysis*, p. 1, Cambridge University Press, Cambridge, 1995

Straathof, A., Panke, S., Schmid, A., "The production of fine chemicals by biotransformations", *Curr. Opin. Biotechnol.*, *13*, 548–556, 2002.

Sumner, J.B. and Somers, G. F., *Chemistry and Methods of Enzymes*, Academic Press, New York, XIII-XVI, (1953)

Tauber, H., *The Chemistry and Technology of Enzymes*. Wiley, New York, (1949)

Uhlig, H., *Industrial enzymes and their applications*, John Wiley & Sons, Inc., New York, (1998)

WCED – World Commission on Environment and Development, *Our common future*, Oxford Univ. Press, Oxford, (1987)

Whitaker, J.R., Voragen, A.G.J., Wong, D.W.S. (Eds.), *Handbook of food enzymology*, Marcel Dekker, New York, (2003)

Index